高等职业教育自动化类新形态一体化教材

C语言程序设计项目双语教程

主编 何玲 罗欢 刘倍雄

中国水利水电出版社
www.waterpub.com.cn
·北京·

内 容 简 介

本书从实际应用出发,将学生成绩管理系统进行分解,形成6个可操作性的任务,主要学习顺序结构程序设计、选择结构程序设计、循环结构程序设计、数组操作、函数模块化程序设计等相关知识,编写时充分考虑高职高专学生的学情,从编程基础能力培养出发,注重知识技能的应用性、工程项目的实践性,以学生熟悉的实际项目制作展开,深入浅出,通过对本书的学习,使学生能更好地掌握C语言编程知识和典型的项目实践应用。本书采用中英文对照混排,既方便初学者熟悉相关概念和内容,也便于英文非母语的读者熟悉英文专业词汇。

本书可作为高等院校计算机、信息技术等大类专业群C语言程序设计双语教材,也可供程序员和编程爱好者参考使用。

图书在版编目(CIP)数据

C语言程序设计项目双语教程:汉文、英文 / 何玲,罗欢,刘倍雄主编. -- 北京:中国水利水电出版社,2023.8
高等职业教育自动化类新形态一体化教材
ISBN 978-7-5226-1578-3

Ⅰ. ①C… Ⅱ. ①何… ②罗… ③刘… Ⅲ. ①C语言-程序设计-高等职业教育-教材-汉、英 Ⅳ.①TP312.8

中国国家版本馆CIP数据核字(2023)第115699号

书　　名	高等职业教育自动化类新形态一体化教材 **C语言程序设计项目双语教程** C YUYAN CHENGXU SHEJI XIANGMU SHUANGYU JIAOCHENG
作　　者	主编　何　玲　罗　欢　刘倍雄
出版发行	中国水利水电出版社 (北京市海淀区玉渊潭南路1号D座　100038) 网址:www.waterpub.com.cn E-mail:sales@mwr.gov.cn 电话:(010)68545888(营销中心)
经　　售	北京科水图书销售有限公司 电话:(010)68545874、63202643 全国各地新华书店和相关出版物销售网点
排　　版	中国水利水电出版社微机排版中心
印　　刷	北京印匠彩色印刷有限公司
规　　格	184mm×260mm　16开本　9.25印张　252千字
版　　次	2023年8月第1版　2023年8月第1次印刷
印　　数	0001—1000册
定　　价	**32.00元**

凡购买我社图书,如有缺页、倒页、脱页的,本社营销中心负责调换
版权所有·侵权必究

Abstract

This tutorial integrates the study of C language with real-world applications, centering around a student performance management system. Upholding the principle of "necessary and sufficient", it places emphasis on practical engineering experience and the development of students' problem-solving and hands-on abilities.

Drawing from years of project-based teaching reform and practice, the authors have crafted this tutorial in accordance with the essential requirements of vocational and technical education. By deconstructing the student performance management system into six manageable engineering projects and tasks, the tutorial covers a range of programming aspects such as sequential structure design, selection structure design, loop structure design, array operations, and function modularization. The content is meticulously tailored to vocational and technical students, focusing on the applicability of knowledge and skills, the practical nature of engineering projects, and the use of familiar real-world scenarios. Through a clear and engaging approach, this tutorial aims to help students gain a better grasp of C language programming concepts and excel in practical applications of typical projects.

All examples in this tutorial have been thoughtfully chosen for their relevance to daily life, with a strong emphasis on hands-on participation. Each example provides beginners with problem-solving guidance, while the targeted exercises aim to enhance readers' interest in programming and improve their coding abilities. The book features a bilingual Chinese-English layout, making it convenient for beginners to acquaint themselves with relevant concepts and content, while also assisting non-native English speakers in learning professional vocabulary.

This tutorial is suitable as a bilingual textbook for higher education institutions teaching computer science, information technology, and other related fields of major professional group C language programming, as well as a valuable reference for programmers and programming enthusiasts.

前 言

在众多的程序设计语言中，C 语言以其灵活性和实用性成为目前使用最广泛的高级程序设计语言之一，几乎任何一种机型、任何一种操作系统都支持 C 语言开发。C 语言程序支持大型数据库开发和 Internet 应用，其应用领域在不断拓展。因此，C 语言程序设计成为工科专业必修的专业基础课程。

本教材以 C 语言为工具，打破传统的教材体系，改以工作任务为载体，以工作过程（即程序设计的过程）为依据，整合、简化教学内容，科学设计学习性工作任务。使学生掌握程序设计的基本思想、方法和技术内涵，着重培养学生分析问题、解决问题的能力，同时，让学生在学习程序设计的过程中，养成良好的编程习惯和编程风格，为后续的专业应用性课程和系统开发课程的学习打下良好的基础。

在内容选取时，注重针对性和适用性相结合，以实现课程目标为依据，以提高学生程序设计能力为核心，以应用性项目开发为主线，以 C 语言语法和结构为基础，以工作任务（学习任务）为载体，设计综合性的学习任务（项目）。本教材以完成"学生成绩管理系统"为主线，任务的开发实施能将课程的全部内容具体化。在研究和分析完成该任务所需要的知识结构的基础上，将课程内容进行重构，分为 6 个主任务，并且把主任务细分为了 15 个相对独立的子任务，每一个子任务包含一个完整的工作过程，子任务之间有相对的独立性，同时遵循知识的连续性。各个任务的主要内容如下。

任务 1：学生成绩总分和平均分的计算。通过任务的实施了解 C 语言程序的开发环境，C 语言的数据类型以及输入/输出函数、算术运算表达式等。

任务 2：学生成绩等级转换。通过任务的实施熟悉 C 语言的关系运算符和运算表达式、逻辑运算符和逻辑表达式的使用；熟练掌握 if 语句和 switch 语句的程序设计。

任务 3：学生成绩分组汇总。通过任务实施理解循环结构程序的设计，熟练掌握 for、while、do-while 语句的程序设计，掌握转移控制语句 break、continue 语句的使用。

任务 4：学生成绩排序。通过任务实施掌握数组的定义、存储结构、输入/输出和使用方法。

任务 5：学生成绩分类汇总。通过任务实施理解模块化设计的思想，学会程序的模块化设计，熟悉形式参数与实际参数的概念，掌握函数的定义和调用、函数的类型和返回值。

任务 6：学生成绩单制作。通过任务实施系统掌握结构体类型、结构体变量的定义、成员访问、结构体数组的定义和使用。

本教材采用中英文对照，既方便初学者熟悉相关概念和内容，也便于英文非母语的读者熟悉英文专业词汇。

本教材由广东水利电力职业技术学院的何玲、罗欢、刘倍雄、Michael James Shipley 等老师组织编写，由何玲负责全书的统稿。由于作者水平有限，书中缺点和不足之处在所难免，恳请读者批评指正。

编者

2023 年 4 月

Preface

Among the plethora of programming languages, the C language stands out due to its flexibility and practicality, making it one of the most extensively used high-level programming languages. Almost every computer model and operating system supports C language development. C language programs facilitate the development of large-scale databases and Internet applications, with its application domain constantly expanding. Consequently, C language programming has become an essential foundational course for engineering majors.

This book employs the C language as its primary tool, breaking away from traditional textbook frameworks and adopting a task-based approach. It is structured around the work process (i.e., the process of program design), integrating and streamlining teaching content while scientifically designing learning-oriented work tasks. Our objective is to enable students to comprehend the fundamental concepts, methodologies, and technical nuances of program design, emphasizing the cultivation of students' analytical and problem-solving abilities. Simultaneously, the book encourages students to develop commendable programming habits and styles throughout the learning process, laying a robust foundation for subsequent application-oriented courses and system development courses.

When selecting content, we prioritize a blend of specificity and applicability, basing our approach on the achievement of course objectives, the enhancement of students' programming capabilities at its core, application-oriented project development as the main thread, C language syntax and structure as the foundation, and work tasks (learning tasks) as the carrier. We have designed comprehensive learning tasks (projects) accordingly. This textbook revolves around the completion of a "Student Performance Management System" project, which embodies the entire courses content through task development and implementation. Building on the research and analysis of the knowledge structure required to complete this task, divided into 6 main tasks, and the main task is subdivided into 15 relatively independent sub-tasks. Each subtask encompasses a complete work process and maintains relative independence between subtasks while adhering to the continuity of knowledge. The main contents of each task are outlined below.

First Task: The Calculation of Total and Average Score for Students. Through the implementation of this task, students will become acquainted with the C language programming development environment, data types, input/output functions, arithmetic expressions, and related concepts.

Second Task: Grade Transformation of Student Score. By executing this task, students will familiarize themselves with the usage of C language relational operators and expressions, logical operators and expressions, and gain proficiency in designing programs with if and switch statements.

Third Task: The Grouping and Summarizing of Student Scores. Through the execution of this task, students will comprehend the design principles of loop structure programs, mastering the program design of for, while, and do-while statements, along with the appropriate use of transfer control statements break and continue.

Fourth Task: Sorting student scores. By completing this task, students will grasp the concepts of array definition, storage structure, input/output, and application methods.

Fifth Task: Categorizing and summarizing student scores. Through the execution of this task, students will gain an understanding of the modular design approach, and learn to create modular programs. They will become familiar with the concepts of formal parameters and actual parameters, mastering the definition, invocation, types, and return values of functions.

Sixth Task: Producing student report cards. By implementing this task, students will acquire proficiency in defining structure data types, structure variables, member access, as well as defining and utilizing structure arrays.

This book employs a bilingual format, featuring both Chinese and English text. This approach not only facilitates beginners in familiarizing themselves with relevant concepts and content but also aids non-native English speakers in acquainting themselves with specialized English vocabulary.

This book is an ideal textbook for students enrolled in bilingual C language courses and can also serve as supplementary material for those using foreign original C language textbooks. Additionally, it is a valuable resource for other students and professionals who wish to learn related professional English.

A team from Guangdong Water Resources and Electric Power Vocational and Technical College, including teachers He Ling, Luo Huan, Liu Beixiong, and Michael James Shipley compiled this book, with He Ling overseeing the manuscript. Throughout the writing process, the authors consulted numerous books and materials related to the C language and would like to express their gratitude to the authors of these reference works. Due to the limited expertise of the authors, shortcomings and deficiencies within the book are inevitable. We sincerely welcome readers to provide constructive critiques and suggestions for improvement.

Authors
August 2023

目 录

前言　Preface

1　　任务 1　学生成绩总分和平均分的计算
　　　　First Task: The Calculation of Total and Average Score for Students

22　任务 2　学生成绩等级转换
　　　　Second Task: Grade Transformation of Student Score

47　任务 3　学生成绩分组汇总
　　　　Third Task: The Grouping and Summarizing of Student Scores

63　任务 4　学生成绩排序
　　　　Fourth Task: Sorting Student Scores

86　任务 5　学生成绩分类汇总
　　　　Fifth Task: Categorizing and Summarizing Students Scores

113　任务 6　学生成绩单制作
　　　　Sixth Task: Producing Students Report Cards

130　附录 A　运算符的优先级和结合性
　　　　Appendix A　Priority and Associativity of Operators

132　附录 B　ASCII 字符编码表
　　　　Appendix B　ASCII Character Encoding Table

134　附录 C　常用标准库函数
　　　　Appendix C　Common Standard Library Functions

任务 1
学生成绩总分和平均分的计算

First Task: The Calculation of Total and Average Score for Students

【知识目标】
1. 理解 C 语言的数据类型。
2. 掌握输入/输出函数。
3. 掌握算术运算表达式。

【能力目标】
1. DEV-C++ 的使用。
2. 程序的编译和调试。

【重点、难点】
1. 输入/输出函数。
2. 运算符和运算表达式。

【课程思政】
通过学习输入/输出函数的格式要求,培养学生养成认真务实、做事遵守规则,做一个遵纪守法的好公民。

【推荐教学方法】
通过教学做一体化教学,结合 DEV-C++ 使学生掌握知识点并学会编写调试程序。

【推荐学习方法】
通过完成任务,在做中学、学中做,掌握实际技能与相关知识点。

【Knowledge Objective】
1. Comprehending the data type of C language.
2. Mastering input and output statement.
3. Mastering the operation expression.

【Competency Objective】
1. The utilization of DEV-C++.
2. The compiling and debugging of programs.

【Focal and Difficult Points】
    1. Inputting and Outputting function.
    2. Operator and operation expression.

【Curriculum Ideological and Political Education】   By means of learning the format requirements of inputting and outputting statements, cultivating the students to be serious and pragmatic, obeying the rules, becoming a law-abiding good citizen.
【Recommended Teaching Method】   By means of integrated teaching, combining the DEV-C++ to make students master knowledge points and learn to compile and debug programs.
【Recommended Learning Method】   By means of accomplishing tasks, learning by doing, doing by learning, having a good command of actual technologies and related knowledge points.

1.1 任务描述

某班级进行了一次 C 语言程序设计考试，请输入学生成绩，并求出他们的总分及平均分。

1.2 相关知识

1.2.1 C 语言程序运行环境及运行方法

C 语言程序必须经过编辑、编译、链接的过程才能生成一个可执行程序，运行 C 语言程序的环境很多，本书主要采用 DEV-C++。

（1）打开 DEV-C++，界面如图 1.1 所示。

单击【文件】→【新建】→【源代码】，空白工作区域就可以输入程序，如图 1.2 所示，当程序输入完毕后，单击▦按钮，或者按 Fn+F11 进行编译运行，此时需要

1.1 Task Description

The examination regarding C Language Program Design was carried out in certain class, please input the student scores, calculating their total and average scores.

1.2 Relevant Knowledge

1.2.1 Runtime Environment and Method of C Language Program

The operational program of C language can only be produced by means of editing, compiling and linking, there are much environment for operating C language program, the DEV-C++ was primarily applied in this work.

(1) Opening DEV-C++, the interface was presented in Figure 1.1.

Clicking【File】→【New】→【Source file】, program can be input in the blank working region, as it was presented in Figure 1.2, while accomplishing inputting the program, clicking the button of ▦, or

First Task: The Calculation of Total and Average Score for Students

图 1.1 DEV-C++ 界面
Figure 1.1 The Interface of DEV-C++

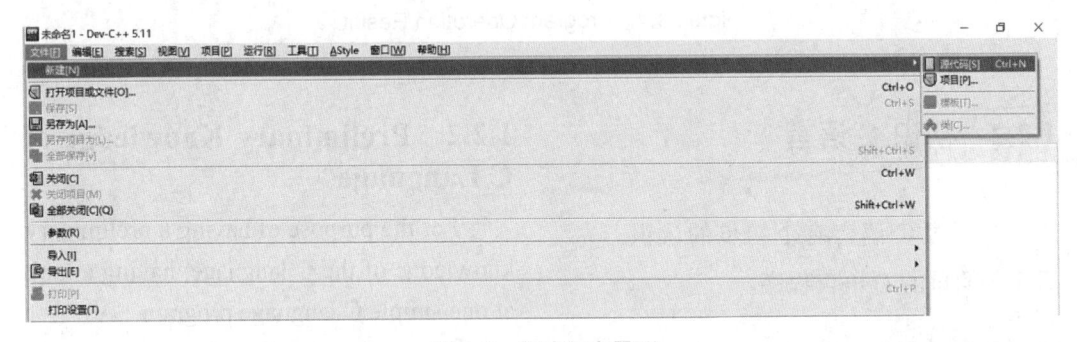

图 1.2 新建程序界面
Figure 1.2 Newly Constructed Program Interface

选择程序保存的位置，并输入文件名，然后单击保存，如图 1.3 所示，如果程序没有错误，即可得出程序运行的结果，如图 1.4 所示。

pressing Fn+F11 to compile and operate, at this moment, it is necessary to select the saving position of programs, and inputting the document name, then clicking to save, as it was presented in Figure 1.3, if there is no error for programs, the operation result of programs can be obtained, as it was presented in Figure 1.4.

图 1.3 保存界面
Figure 1.3 Saving Interface

图 1.4 程序运行结果
Figure 1.4 Program Operation Result

1.2.2 初识 C 语言

为了对 C 语言有个初步的认识，先看一个简单的 C 语言程序。

【例 1.1】在屏幕上输出一行文本信息"Hello World!"。

【解题思路】直接使用 printf() 函数的输出功能即可实现。

1. 预处理命令（包含头文件）

#include 将 < stdio.h > 或 "stdio.h" 文件包括到用户源文件中。即 # include <stdio.h> 或 # include "stdio.h"

1.2.2 Preliminary Knowledge of C Language

For the purpose of having a preliminary knowledge of the C language, having a look at one simple C language program.

【Example 1.1】Output one line of text message "Hello World!" on the screen.

【Problem-Solving Ideas】It can be realized directly by utilizing the output function of printf () function.

1. Preprocessing Order (Including Header Files)

#include incorporates the file of < stdio.h > or "stdio.h" into the user source file. Namely # include <stdio.h>or # include "stdio.h"

First Task: The Calculation of Total and Average Score for Students

stdio.h 包含了与标准 I/O 库有关的变量定义和宏定义。在需要使用标准 I/O 库中的函数时，应在程序前使用上述预编译命令，预编译命令要写在程序的最开头。

2. main() 函数

C 程序是一种函数结构，一般有一个或若干个函数组成，其中有且只有一个名为 main() 的主函数，程序的执行从这里开始，且不论其在程序的什么位置。

3. 函数的组成

C 语言函数由函数首部和函数体两部分组成。

（1）函数首部是函数的第一行，一般包括函数类型、函数名、原括号和函数参数（可以缺省），如 int main()。

（2）函数体是函数首部下一对"{ }"括起来的部分。函数体一般包括声明部分（定义本函数所使用的变量）和执行部分（由若干条语句组成的命令序列）。

4. 程序书写格式

（1）所有语句都必须以";"结束，表示语句的结束。

（2）程序行的书写格式自由，1 行可以写多条语句，1 条语句也可以分多行书写。

stdio.h includes the variable and macro definitions regarding standard I/O database. When there is a need for utilizing the functions in standard I/O database, the above precompiling order should be utilized before program, the precompiling order should be written in the very beginning of program.

2. main ( ) Function

C program constitutes a kind of function structure, which is composed of one or several functions, thereinto, only one main function of main(), the implementation of programs starts from it, regardless of its position in program.

3. The Composition of Function

The C language function is composed of two parts of function heading and function body.

(1) The function heading refers to the first line of function, including function type, function name, original parenthesis and function parameter (can be omitted), such as int main().

(2) The function body refers to one pair within parenthesis "{ }" under function heading. The function body generally includes statement part (the variables utilized to define this function) and implementation part (the order sequence composed of several statements).

4. Program Writing Form

(1) All of the statements must be ended with ";", signifying the ending of statements.

(2) The writing form of program line is free, one line can be written with several statements, one statement can be divided into several lines.

5. 注释

可以用"/*…*/"和"//"对 C 语言程序中的任何部分进行注释。注释内容不会被编译器编译。注释可以提高程序的可读性，使用注释是编程人员的良好习惯。

（1）"/*…*/"是块注释,可注释多行,"/*"和"*/"必须成对出现将注释内容括起来,且"/"和"*",以及"*"和"/"之间不能有空格,否则会出错。

（2）"//"是行注释,只注释当前行。

1.2.3 数据类型

在 C 语言中，数据类型指的是用于声明不同类型的变量或函数的一个广泛的系统。变量的类型决定了变量存储占用的空间，以及如何解释存储的位模式。C 语言的数据类型见表 1.1。

C 语言程序中的数据有常量和变量之分。

1. 常量

在程序运行中，其值不能被改变的量称为常量。常量可分为：整型常量、浮点型常量、字符常量、字符串常量、符号常量等。

整型常量,例：1, 2, –10。

5. Annotation

"/*…*/" and "//" can be applied to annotate any parts in C language program. The annotation content cannot be edited by compiler. Annotation can improve the readability of program, and it is a good habit for programming personnel to utilize annotations.

(1) "/*…*/" is annotation, several lines can be annotated, "/*" and "*/" must appear in pairs to include the annotation content, no blank can be found between "/" and "*" as well as "*" and "/", or else a mistake will be made.

(2) "//" is a line of annotation, which can only be utilized to annotate the present line.

1.2.3 Data Type

Data type refers to an extensive system utilized to state the variables of various types or functions in C language, the type of function defines the space occupied by variable storage, as well as the bit pattern utilized for explaining storage. The type of C language is presented in Table 1.1.

The data in C language program can be classified into constant quantity and variable.

1. Constant Quantity

The unchangeable quantity operated in program refers to constant quantity. The constant quantity can be classified as integer constant, floating-point type constant, character constant, character string constant, symbol constant and so on.

Integer constant, such as 1, 2, –10.

First Task: The Calculation of Total and Average Score for Students

表 1.1　C 语言的数据类型

序号	类型和描述
1	基本类型：算术类型，包括整数类型和浮点类型两种类型
2	枚举类型：算术类型，被用来定义在程序中只能赋予其一定的离散整数值的变量
3	void 类型：类型说明符 void 表明没有可用的值
4	派生类型：包括指针类型、数组类型、结构类型、共用体类型和函数类型

Table 1.1　Data Type of C Language

Number	Type and Description
1	Basic Type: It is arithmetic type, including two types of type integer and floating-point type
2	Enumeration Type: It is also arithmetic type, which is utilized to define the variable that can only be endowed with certain discrete integral value in program
3	void Type: The type specifier void implied that there is no useful value
4	Derivation Type: It includes pointer type, array type, structure type, union type and function type

浮点型常量由整数部分、小数点、小数部分和指数部分组成。例：1.23，–9.8，.123，23.，0.0，2.23e-4（表示 2.23×10^{-4}），0.23e-3（表示 0.23×10^{-3}），1.23e3（表示 1.23×10^{3}）。

字符常量是括在单引号中，例如，'x' 可以存储在 char 类型的简单变量中。例：'A'，'+'，'8'。

字符串常量是括在双引号中的，例如，"hello"，"A"。

符号常量，例 #define PI 3.14，其中 PI 是一个符号常量，其值为 3.14，它不能在程序中被改变。

2. 变量

变量就是在程序运行过程中，其值可以被改变的量。一个变量由两个要素组成，即变量名和变量值。

Floating-point type constant is composed of integer part, decimal point, decimal part and exponent part. Such as 1.23, -9.8, .123, 23., 0.0, 2.23e-4 (signifying 2.23×10^{-4}), 0.23e-3(signifying 0.23×10^{-3}), 1.23e3(signifying 1.23×10^{3}).

Character constant is included in single quotes, for example, 'x' can be stored in simple constant of char type. Such as 'A', '+', '8'.

Character string constant is included in double quotation marks, such as "hello", "A".

Symbol constant, such as #define PI 3.14. While PI is a symbol constant, its value is 3.14, which cannot be changed in program.

2.Variable

The so-called variable refers to the changeable quantity in operating program. One variable is composed of two elements. Namely, variable name and value.

每一个变量都必须有一个名字，即变量名。变量名的命名规则：由字母或下划线开头，后面跟字母、数字和下划线。其有效长度，随系统而异，但至少前 8 个字符有效。如果超长，则超长部分会被舍弃。

注意：

（1）C 语言的关键字不能用做变量名。

（2）C 语言的变量名区分大小写，即同一字母的大小写，都被认为是两个不同的变量。例 Total、total、toTal 是不同的变量名。

（3）给变量名命名时，最好遵循"见名知意"这一基本原则。例如，name/xm(姓名)、sex/xb(性别)、age/nl(年龄)、salary/gz(工资)。

3. 变量的定义及初始化

变量的定义是指定一个数据类型，并包含了该类型的一个或多个变量的列表。

变量定义的一般格式：

类型说明符 变量1，变量2，……；

其中，类型说明符是 C 语言中的一个有效的数据类型，如整型类型说明符 int、字符型类型说明符 char 等。

例如：

```
int a,b,c;
char cc;
```

Each variable must have its name, that is the variable name. The naming rule for variable: started with letters or underlines, followed with letters, numbers and underlines. Its effective length can vary with system, but at least eight characters are effective. If it is overlength, then the overlength part will be abandoned.

Attention:

(1)The keywords of C language cannot be used as variable name.

(2)The uppercase and lowercase forms are distinguished for variable names of C language, namely, the uppercase and lowercase forms of same letters are both considered as two different variables. For example, Total, total, toTal belong to different variable names.

(3) While naming variables, it is better to observe the basic rule of "recognizing the meaning through knowing the name". Such as name/xm (name), sex/xb (gender), age/nl (age), salary/gz (salary).

3.The Definition and Initialization of Variable

The definition of variable refers to designate one data type, which including one or several variables lists of this type.

The general form of variable definition:

Type specifier variable 1, variable 2, ...;

Thereinto, type specifier constitutes one effective data type in C language, such as the integer type specifier int, the character type specifier char and so on.

Such as

```
int a,b,c;
char cc;
```

In C language, it is required to define all the available variables compulsively, that is to "first define, then utilize".

The general form for variable initialization:

Data type Variable name [=initial value, variable name 2=initial value 2, …];

For example

```
int x=1,y=2,z=3;
//Defining integer type variables
float a=1.1,b=1.2,c=-0.1;
//Defining floating-point type variables
char ch1='A',ch2='*';
//Defining character type variables
```

1.2.4 Format Output Function—printf () Function

Format output printf () function, its function is to output several data of random types to display. The general form of printf () function: printf ("format control character strings"[,output list]); For example printf ("%d, %c\n", I, c).

Format control character strings: it is a character string included by double quotation marks, including two kind of information:

(1) The format character started with %, which can be utilized for the output format of designated data. For example, %d: outputting integers with symbols in the form of decimalism; %c: outputting single character; %x: outputting integers without symbols in the form of hexadecimal system, %s: outputting character strings; %f: outputting real number in decimal form; %e: outputting real number in exponential form.

(2) The outputting characters with original form can play a reminder role in displaying.

| 1 | printf("This is my first C program.\n"); | //原样输出一串字符，并换行 |
| 2 | printf("sum=%d\n", s); | //原样输出 sum=，然后以 %d 形式输出变量 s 的值，并换行 |

| 1 | printf("This is my first C program.\n"); | //Outputting a string of characters with original form, and changing lines |
| 2 | printf("sum=%d\n", s); | //Outputting sum= with original form, then output the value of s in the form of %d, and changing lines |

注意：

输出列表是需要输出的一系列参数，其个数必须与格式化字符串所说明的输出参数个数相同，各参数之间用"，"分开，且顺序一一对应，否则将会出现意想不到的错误。

【例 1.2】商场的每个商品都需要打印销售标签，上面包含商品名称、销售价格、包装规格、产地等信息，现有风行牛奶，每盒 4.5 元，每盒容量 250ml，产地广州，生产日期 2021 年 9 月 28 日。请编程序打印一个标签。

【解题思路】定义年月日和容量为整型变量并赋初值，定义价格为浮点型变量并赋初值，然后利用 printf() 函数输出结果即可实现。

Attention:

There is a series of parameters in outputting lists, its number must be consistent with the outputting parameters specified by formatting character strings, the various parameters are divided by "，", and performing one-to-one correspondence, or there will be unexpected error.

【Example 1.2】Sales labels must be printed for every merchandise in shopping mall, including merchandise name, sales price, package specification, origin place and so on, now there is FengXing milk, 4.5 yuan for each box, each box is of 250ml, the origin place is GuangZhou, the production date is September 28th, 2021. Please print a label by programming.

【Problem-Solving Ideas】Defining the date and capacity as integer type variables and assigning initialized value, defining price as floating-point type variable and assigning initialized value, then taking advantage of printf () function to output result.

First Task: The Calculation of Total and Average Score for Students

```c
#include "stdio.h"              //预处理命令
void main( )                    //主函数
{
    int iYear=2021,iMonth=9,iDay=28,iVol=250;   //定义整型变量并赋值
    float fPrice=4.5;                            //定义实型变量并赋值
    printf("Name:FengXing Milk\n");              //输出字符串
    printf("Size:%d ml\n",iVol);                 //输出一个整数
    printf("Price:RMB%0.2f\n",fPrice);           //输出一个实数
    printf("Origin:GuangZhou\n");                //输出字符串
    printf("Date:%d.%d.%d\n",iYear,iMonth,iDay); //输出多个整数
}
```

程序运行结果： Program operation result:

```
Name:FengXing Milk
Size:250 ml
Price:RMB4.50
Origin:GuangZhou
Date:2021.9.28
```

注意：

使用 printf() 函数时，要求格式控制字符串中必须含有与输出项一一对应的格式符，并且类型要匹配。printf() 函数也可以没有输出项。

Attention:

While utilizing printf() function, it is required that the format control character strings should contain format characters corresponded to output items, the type must match. Printf() function can also be without output items.

1.2.5 格式输入函数——scanf() 函数

格式输入函数 scanf() 函数，作用是按用户指定的格式从键盘把数据输入到指定的变量地址中，其一般形式为：scanf(格式控制字符串，地址列表)。

格式控制字符串的作用与 printf() 函数类似。例如：%d 是以十进制形式输入

1.2.5 Format Input Function—scanf () Function

As for format input scanf() function, its function is to input the data from keyboard to designated variable address in accordance with the format proposed by users, its general form is scanf (format control character strings, address list).

The function of format control character strings is similar to that of printf() function.

带符号整数；%c 是输入单个字符；%x 是以十六进制形式输入无符号整数；%s 是输入字符串；%f（%lf）是以实数形式输入单精度（双精度）实数。

地址列表是由若干个地址组成的列表，可以是变量的地址、数组的首地址、字符串的首地址。变量的地址是由地址运算符"&"后跟变量名组成的。多个地址之间要用逗号隔开。

【例 1.3】用 scanf() 函数输入多个数值数据。

【解题思路】定义变量并利用 scanf() 进行多行输入，利用 printf() 进行多行输出即可实现。

For example %d means to input integers with symbols in decimal form; %c means to input single character; %x means to input integers without symbols in hexadecimal system; %s means to input character strings; %f (%lf) means to input real numbers with single precision (double precision) in the form of real number.

Address list is a list composed of several addresses, which can be the address of variables, the initial address of arrays, and the initial address of character strings. The address of variables is composed of the address operation character "&" followed by variable names. Several addresses can be divided by comma.

【Example 1.3】Taking advantage of scanf () function to input several value data.

【Problem-Solving Ideas】Defining variables and taking advantage of scanf () to carry out the input for several lines, it can be realized by utilizing printf () to perform input for several lines.

```
例1-3.c
1   #include<stdio.h>                              //预处理指令
2   void main(  )                                  //头文件
3   {
4       int i,j;                                   //定义两个整型变量
5       float k;                                   //定义一个浮点型变量
6       double x;                                  //定义一个双精度变量
7       printf("请输入两个整数和两个实数：\n");      //输出提示语句
8       scanf("%d %d %f %lf",&i,&j,&k,&x);         //变量输入语句
9       printf("%d,%d,%f,%lf\n",i,j,k,x);          //变量输出语句
10  }
```

&i, &j, &k, &x 中的 "&" 是 "取地址运算符"，&i 表示变量 i 在内存中的地址。上面 scanf 函数的作用时将输入的

The "&" in &i, &j, &k and &x is "address-of operator", &i means the address of variable i in internal storage. The function

First Task: The Calculation of Total and Average Score for Students

4个数值数据一次存入变量 i、j、k、x 的地址中去。

说明：

（1）用 scanf() 函数一次输入多个数值或多个字符串时，在两个数据之间可用一个（或多个）空格、换行符（按 Enter 键产生换行符）或 Tab 符（按 Tab 键产生 Tab 符）作分割。换言之，用 scanf 函数输入数据时，系统以空格、换行符或 Tab 符作为一个数值或字符串的结束符。

例如，【例 1.3】的运行情况：

1）用空格作输入数据之间的分隔：

```
请输入两个整数和两个实数：
45 67 8.9 12.345689
45, 67, 8.900000, 12.345689
```

2）用换行符作输入数据之间的分隔：

```
请输入两个整数和两个实数：
45
67
8.9
12.345689
45, 67, 8.900000, 12.345689
```

3）用 Tab 符作输入数据之间的分隔：

```
请输入两个整数和两个实数：
45      67      8.9     12.345689
45, 67, 8.900000, 12.345689
```

（2）当输入数据的类型与 scanf 函数中的格式符指定的类型不一致时，系统认为该数据结束。

注意：

用 "%d %d %f %lf" 格式输入数据时，

of above scanf is to save the input four value date into the address of variables i, j, k and x for one time.

Description:

(1) While utilizing scanf() function to input several values or character strings for one time, one blank (or several ones), line break (pressing Enter button to produce line break) or Tab symbol (pressing Tab button to produce Tab symbol) can be utilized to divide between two data. In other words, while applying scanf function to input data, the blank, line break or Tab symbol are taken as the ending symbols for one value or character strings in system.

For example, the operation condition of 【Example 1.3】 was presented below:

1) Utilizing blank as the separator among data:

```
请输入两个整数和两个实数：
45 67 8.9 12.345689
45, 67, 8.900000, 12.345689
```

2) Applying line break as the separator among data:

```
请输入两个整数和两个实数：
45
67
8.9
12.345689
45, 67, 8.900000, 12.345689
```

3) Utilizing Tab symbol as the separator among data:

```
请输入两个整数和两个实数：
45      67      8.9     12.345689
45, 67, 8.900000, 12.345689
```

(2) While the type of input data is inconsistent with the designated type for format character in scanf function, the system would regard it as the ending for data.

Attention:

While applying "%d %d %f %lf"

不能用逗号作输入数据间的分隔符。若实在想用逗号作输入数据的分隔符，可改为"%d, %d, %f, %lf"格式，但不提倡这样做。

（3）用 scanf() 函数输入字符时，系统将输入的空格、换行符作为有效字符。

【例 1.4】用 scanf() 函数输入多个字符。

format to input data, the comma cannot be used as the separator among inputting data. If one really wants to utilize the comma as separator among inputting data, it can be changed into the formats of "%d, %d %f %lf", but this is not advocated.

(3) While utilizing scanf() function to input characters, the system would regard the input blank or line break as effective characters.

【Example 1.4】Utilizing scanf() function to input several characters.

```
1  #include<stdio.h> //预处理命令：包含输入输出头文件
2  void main()       //主函数
3  {
4      char i,j,k;   //定义三个字符变量
5      printf("请输入三个字符：\n");//提示输入三个字符变量
6      scanf("%c%c%c",&i,&j,&k);//输入三个字符变量分别赋给变量i, j, k
7      printf("%c,%c,%c\n",i,j,k);//将变量i, j, k的数据输出
8  }
```

运行情况：

请思考为什么？

请输入三个字符：
a b c
a, ,b

Operation condition:

Please reflect on the reason?

请输入三个字符：
a b c
a, ,b

1.2.6 字符输出函数——putchar() 函数

putchar() 函数的功能是向显示器输出字符变量 c 对应的字符，其一般格式为：putchar(c)。

【例 1.5】输出一个字符。

1.2.6 Character Output Function— putchar() Function

The function of putchar() function is to input the characters corresponded to character variable c to display, its general format is: putchar (c).

【Example 1.5】Output one character.

```
1  #include <stdio.h>        //包含输入输出库函数头文件
2  void main()               //主函数
3  {
4      char a,b,c;           //定义三个字符变量
5      a='H'; b='X'; c='Y';  //给字符变量分别赋值
6      putchar(a); putchar(b); putchar(c); putchar('\n'); //在屏幕上显示字符
7  }
```

1.2.7 字符输入函数——getchar()函数

getchar() 函数的功能是从键盘上输入一个字符，其一般格式为：getchar()。通常将键盘上输入的字符赋予一个字符变量，构成赋值语句。

【例 1.6】从键盘输入一个字符，并在屏幕上显示。

1.2.7 Character Iutput Function——getchar() Function

The function of getchar() function is to input a character from keyboard, its general format is: getchar(). The input character on keyboard is generally assigned with a character variable to construct assignment statement.

【Example 1.6】Input a character from keyboard, and display it on screen.

```
1  #include <stdio.h>           //包含输入输出库函数头文件
2  void main()                  //主函数
3  {
4      char c;                  //定义一个字符变量c
5      printf("请输入一个字符：");  //提示语句
6      c=getchar();             //从键盘输入一个字符赋值给变量c
7      putchar(c);              //在屏幕上输出c
8      putchar('\n');           //换行
9  }
```

1.2.8 算术运算符及算术运算表达式

1. 五种基本的算术运算符

+（加法）、-（减法）、*（乘法）、/（除法）、%（求余数）

这里，需要特别提出的是：

（1）除法运算/。

C语言规定，两个整数相除，其商为整数，小数部分被舍弃。例如：5/2=2。

1.2.8 Arithmetic Operator and Arithmetic Operation Expression

1. Five Kinds of Basic Arithmetic Operators

+ (Addition), −(subtraction), *(multiplication),/(Division method), % (calculating remainder)

At this point, it is necessary to propose that:

(1)Regarding division operation/.

As it is prescribed by C language, dividing two integers, its quotient is integer, the decimal part will be abandoned. Such

如果商为负数，则取整的方向随系统而异。但大多数的系统采取"向零取整"原则，换句话说，取其整数部分。例如：–5/3=–1。

（2）求余数运算%。

要求两侧的操作数均为整型数据，否则出错。例如：5%3=2，3%5=3，–5%3=–2，–5%（–3）=–2。但是，5.2%3是语法错误。

2. 算术表达式和运算符的优先级与结合性

用算术运算符和括号将运算对象（常量、变量或表达式等）连接起来的、符合C语法规则的式子，称为C算术表达式。例如：a+b*c–5/2+'a'。

C语言规定了运算符的优先级和结合性。在表达式求值时，先按运算符的优先级高低次序执行，例如表达式x–y*z相当于x–(y*z)。如果在一个运算对象两侧的运算符的优先级相同，则按照规定的"结合方向（结合性）"处理。

算术运算符的结合方向为"自左向右（左结合性）"，即先左后右。因此表达式a+b–c相当于(a+b)–c。

【例1.7】将一个两位十进制的整数的

as:5/2=2.

If the quotient is negative, then the direction for rounding numbers will vary with system. But most systems will apply the principle of "rounding zero", in other words, obtaining its integer part. Such as: –5/3=–1.

(2)Regarding calculating remainder operation %.

It is required that the operation numbers in both sides are integer type data, or else a mistake will be made. Such as: 5%3=2, 3%5=3, –5%3=–2, –5%(–3)=–2. However, 5.2%3 is grammar mistake.

2.Arithmetic Expression Coupled with the Priority and Associativity of Operator

Applying arithmetic operators and parenthesis to connect operation subjects (constant quantity, variable or expressions), the formulas that conforming to C language grammar are called C arithmetic expressions. Such as: a+b*c–5/2+'a'.

C language prescribes the priority and associativity of operators. In calculating the values by expressions, first conforming to the priority descending order of operators to perform, for example, the expression x–y*z is equivalent to x–(y*z). If the priority of operators on both sides of operation subjects is the same, then processing according to the prescribed "associativity direction (associativity)".

The associativity direction of arithmetic operators is "left to right (left associativity)", namely, first left and then right. Therefore, the expression a+b–c is equal to (a+b)–c.

【Example 1.7】Separate the tens and

十位数和个位数分离。

【解题思路】对于一个两位十进制的整数 n，其十位为 n/10，个位为 n%0，加入 n 为 98，则十位的 9=98/10，个位 8=98%10，然后利用 printf() 函数输出十位和个位上的数值。

ones of a two digit Decimal integer

【Problem-Solving Ideas】In terms of one two-digit decimal integer n, its tens digit is n/10, units digit is n%0, adding n as 98, then the tens digit is 9=98/10, units digit 8=98%10, then taking advantage of printf () function to output the values on tens digit and unit digit.

```
1  #include <stdio.h>    //包含输入输出库函数头文件
2  void main()           //主函数
3  {
4      char c;                        //定义一个字符变量c
5      printf("请输入一个字符：");     //提示语句
6      c=getchar();                   //从键盘输入一个字符赋值给变量c
7      putchar(c);                    //在屏幕上输出c
8      putchar('\n');                 //换行
9  }
```

1.2.9 强制类型转换运算符及其表达式

强制类型运算转换运算符的作用是将一个表达式转换成所需要的类型，其一般格式为

(类型标识符)(表达式)

例如：

(int) i   将 i 转换为整型
(float)(x+y)   将 x+y 的结果转换为 float 型
(int)x+y   将 x 转换成整型后，再与 y 相加

【例 1.8】将实型数据强制转换为整型。

1.2.9 Type Casting Conversion Operator and Its Expression

The function of type casting conversion operator is to transform one expression into required type, its general format is:

(Type specifier)(expression)

For example

(int) i   Transforming i into integer type
(float)(x+y)   Transforming the result of x+y into float type
(int)x+y   After transforming x into integer type, adding it with y

【Example 1.8】Transforming fact data into integer type compulsively

```
1  #include <stdio.h>         //预处理命令
2  void main()
3  {
4      int i;                 //定义整型变量i
5      float j=3.8;           //定义实型变量j，并赋初值
6      i=(int)j;              //将实型变量j强制转换为int型
7      printf("j=%f,i=%d\n",j,i);//输出数据
8  }
```

程序运行结果：

```
Name:FengXing Milk
Size:250 ml
Price:RMB4.50
Origin:GuangZhou
Date:2021.9.28
```

1.2.10 赋值运算符及其表达式

1. 简单赋值运算符"="及其表达式

简单赋值表达式的形式：变量 = 表达式

赋值表达式"a=5"的值是5。执行运算后，变量 a 的值也是5。

赋值表达式中的"表达式"，又可以是一个赋值表达式。C 语言规定，赋值运算符是按照"自右向左"的结合顺序。

例如：

```
x=5       //将 5 赋给变量 x；
x=6+7     //将 6+7 的值赋给变量 x；
5=x       是错误的；
x+y=z     也是错误的。
```

2. 复合赋值运算符及其表达式

复合赋值运算是 C 语言特有的一种运算。

复合赋值运算的一般格式为

变量 复合运算符 表达式

复合算术运算符有五种，分别是：+=，-=，*=，/=，%=。

1.2.10 Assignment Operator and Its Expression

1. Simple Assignment Operator "=" and Its Expression

The form of simple assignment expression is: variable = expression

The value of assignment expression "a=5" is 5.

After performing operation, the value of variable a is also 5.

The "expression" in assignment expression, which can also be another assignment expression. As it is prescribed by C language, assignment operator should follow the "right to left" associativity order.

For example:

```
x=5       //Assigning 5 to variable x;
x=6+7     //Assigning the value of 6+7 to variable x;
5=x       It is false;
x+y=z     It is also false.
```

2. Compound Assignment Operator and Its Expression

Compound assignment operation is unique to C language.

The general format of compound assignment operation is:

Variable Compound Operation Symbol Expression

There are five kinds of operators for compound assignment operation, they are: +=, -=, *=, /=, %=.

x+=3	等价于 x=x+3	x+=3	Equivalent to x=x+3
x+=5+8	等价于 x=x+(5+8)	x+=5+8	Equivalent to x=x+(5+8)
x*=y+2	等价于 x=x*(y+2)	x*=y+2	Equivalent to x=x*(y+2)
x/=x+y	等价于 x=x/(x+y)	x/=x+y	Equivalent to x=x/(x+y)
x/=8	等价于 x=x/8	x/=8	Equivalent to x=x/8
x%=7	等价于 x=x%7	x%=7	Equivalent to x=x%7
x%=(4−2)	等价于 x=x%(4−2)	x%=(4−2)	Equivalent to x=x%(4−2)

1.3 任务实现

【任务要求】某班级进行了一次 C 语言程序设计考试，请输入学生成绩，并求出他们的总分及平均分。

【任务分析】把三个学生的成绩定义为整型变量，利用 scanf() 函数将三个人的成绩输入，把三个学生的成绩加起来即可求出三个人的总分，然后求出平均分，利用 printf() 函数将三个人的成绩、总分和平均分进行输出即可实现该任务。

1.3.1 学生成绩的输入和输出

把三个学生的成绩定义为整形变量，利用 scanf() 函数将三个人的成绩输入，利用 printf() 函数将三个人的成绩进行输出即可实现。

1.3 Task Implementation

【Task Requirement】The examination regarding C Language Program Design was carried out in certain class, please input the student scores, calculating their total and average scores.

【Task Analysis】Defining the scores of three students as integer variables, taking advantage of scanf() function to input the scores of three students, adding their respective scores to calculate the total scores, then calculating the average score, utilizing printf() function to input the scores, total scores as well average scores of three students so as to realize this task.

1.3.1 The Input and Output of Student Scores

Defining the scores of three students as integer variables, taking advantage of scanf() function to input the scores of three students, utilizing printf() function to input the scores of three students so as to realize this task.

```c
1  #include "stdio.h"                         //文件预处理
2  void main(    )                            //主函数
3  {                                          //函数体开始
4      int x,y,z;                             //定义三个变量x,y,z
5      printf("请输入三个学生的成绩");        //提示语句
6      scanf("%d%d%d",&x,&y,&z);              //输入三个学生的成绩
7      printf("输出三个学生的成绩");          //提示语句
8      printf("x=%d,y=%d,z=%d\n",x,y,z);      //输出三个变量x,y,z的值
9  }
```

程序执行结果：

Program implementation result:

```
请输入三个学生的成绩90 98 78
输出三个学生的成绩x=90,y=98,z=78
```

在这个任务里面主要练习了输入和输出函数的使用。

The application of inputting and outputting functions is primarily practiced in this task.

1.3.2 学生成绩总分和平均分的计算

1.3.2 The Calculation of Total and Average Scores of Students

将输入的三个学生的成绩加起来即可求出三个人的总分，然后再除以 3 即可求出三个人的平均成绩，利用 printf() 函数将三个人的总分和平均分输出即可实现。

Adding the inputting scores of three students to calculate their total scores, then dividing it into 3 to calculate their average scores, utilizing printf() to input the total and average scores of three students so as to realize this task.

```c
1   #include "stdio.h"                                //文件预处理
2   void main(    )                                   //主函数
3   {
4       int x,y,z;                                    //定义三个整型变量
5       float sum,avg;                                //定义二个实型变量sum,avg
6       printf("请输入三个学生的成绩");               //提示语句
7       scanf("%d%d%d",&x,&y,&z);                     //输入三个学生的成绩
8       sum=x+y+z;                                    //将x+y+z的值赋给sum
9       avg=sum/3;                                    //将sum/3的值赋给avg
10      printf("请输出三个学生的总成绩及平均分为");   //输出提示
11      printf("sum=%.2f,avg=%.2f\n",sum,avg);        //输出二个变量sum及avg的值
12  }
```

程序执行结果：

Program implementation result:

```
请输入三个学生的成绩90 98 76
请输出三个学生的总成绩及平均分为sum=264.00,avg=88.00
```

First Task: The Calculation of Total and Average Score for Students

在这个任务里面主要运用了输入/输出和算术运算的相关知识。

【练习与提高】

1. 输入三角形三边的长，编程求三角形的周长及面积。

2. 从键盘输入一个小写字母，要求用大小写字母形式输出该字母及对应的ASCII码值。

3. 假设m是一个三位数，请编程输出由m的个位、十位、百位反序而成的三位数，例如123反序为321。

4. 已知"int x=20，y=32"，编程实现将x和y的值互换。

The knowledge regarding inputting and outputting coupled with arithmetic operation is applied primarily in this task.

【Practice and Improvement】

1. Inputting the length of three sides for triangle, programming to calculate the perimeter and area of triangle.

2. Inputting one small letter from keyboard, it is required to output this letter in uppercase and lowercase forms and the corresponded ASCII code value.

3. Supposing that m is a three-digit number, please program and output the three-digit number composed of the units digit, tens digit and hundreds digit of m in descending order, for example, the descending order of 123 is 321.

4. Given that "int x=20, y=32", programming to realize the conversion between x and y.

任务 2
学生成绩等级转换

Second Task: Grade Transformation of Student Score

【知识目标】
1. 正确使用关系运算符、逻辑运算符、条件运算符。
2. 熟练使用 if、switch 语句的使用方法。
3. 掌握选择语句的嵌套。

【能力目标】
能够熟练地编写分支程序。

【重点、难点】
1. 关系运算符、逻辑运算符的使用。
2. 分支结构的嵌套。

【课程思政】
1. 通过程序流程图的讲解，引导学生做一个做事条理分明的人，能够按照事情的顺序和计划进行，学会统筹管理，提高做事效率。
2. 通过运算符优先级的学习，明白事情有轻重缓急之分，学会先处理重要的和紧急的事情。

【推荐教学方法】
通过教学做一体化教学，结合生活中常见的事例，使学生掌握知识点，学会编制程序流程图并进行程序的编写。

【推荐学习方法】
通过完成任务，在做中学、学中做，掌握实际技能与相关知识点。

【Knowledge Objective】
1. Applying relational operator, logic operator and conditional operator correctly.
2. Familiar with the usage methods of if and switch.
3. Mastering the nesting of select statement.

【Competency Objective】
Proficient in compiling branch program.

【Focal and Difficult Points】
1. The application of relational operator and logic operator.
2. The nesting of branch structures.

Second Task: Grade Transformation of Student Score

【Curriculum Ideological and Political Education】
1. By means of illustration for program flow diagram, guiding the students to be a logical person, being able to do things according to the sequence and plan, learning overall management, and improving work efficiency.
2. By means of priority learning for operator, comprehending the order of importance and emergency, learning to deal with important and urgent things first.

【Recommended Teaching Method】
By means of integrated learning, combining the common cases, making students grasp the knowledge points, learning to compile program flow diagram and edit the program.

【Recommended Learning Method】
By means of accomplishing tasks, learning by doing, doing by learning, grasping the relevant knowledge points regarding actual technologies.

2.1 任务描述

某班级进行了一次 C 语言程序设计考试，教师按百分制给出了学生成绩，现在学校要求改为五级制进行打分，即 90~100 分为 A，80~89 分为 B，70~79 分为 C，60~69 分为 D，60 分以下为 E。分数可以任意输入。

2.2 相关知识

2.2.1 关系运算符与关系表达式

在 C 语言中，关系运算就是做比较，使用关系运算符对两个操作数进行数值大小上的比较，判断其结果是真还是假，从而得到表达式的值为 1 或 0。

2.1 Task Description

One C Language Program Design examination was carried out in certain class, the student scores were issued by teacher based on hundred-mark system, at present, it is required by the school to change it into the five-class system for marking, namely, 90 to 100 points is equivalent to A, 80 to 89 points is equivalent to B, 70 to 79 points is equivalent to C, 60 to 69 points is equivalent to D, below 60 points is equivalent to E. The scores can be input randomly.

2.2 Relevant Knowledge

2.2.1 Relational Operator and Relational Expression

In C language, relational operation means to make comparison, applying relational operator to compare the values of two operation numbers, judging the authenticity of result, then obtaining the value of expression as 1 or 0.

1. 关系运算符

在 C 语言中，有六种关系运算符，见表 2.1。

1.Relational Operator

In C language, there are six kinds of relational operators, the relational operators were presented in Table 2.1.

表 2.1　关 系 运 算 符

序号	运算符	对应的数学运算符	含义	优先级
1	<	<	小于	优先级相同（高）
2	<=	≤	小于等于	
3	>	>	大于	
4	>=	≥	大于等于	
5	==	=	等于	优先级相同（低）
6	!=	≠	不等于	

Table 2.1　Relational Operator

Number	Operator	Corresponded Mathematical Operator	Meaning	Priority
1	<	<	Less than	Same priority (high)
2	<=	≤	Less than or equal to	
3	>	>	More than	
4	>=	≥	More than or equal to	
5	==	=	Equal to	Same priority (low)
6	!=	≠	Not equal to	

（1）运算数：所有的关系运算符都是二元运算符，需要两个操作数参与运算。

（2）结合性：所有的关系运算符都是左结合，即运算时从左向右进行运算。

（3）优先级：前四种关系运算符的优先级相同，后两种关系运算符的优先级相同，前四种的优先级高于后两种的优先级。

(1) Operation Number: All of the relational operators belong to binary operators, two operation numbers are required to participate in the operation.

(2) Associativity: All of the relational operators are left-associative, that is to operate from left to right.

(3) Priority: The former four relational operators have the same priority, the latter two relational operators have the same priority, the priority of former four is higher

关系运算符的优先级低于算术运算符，高于逻辑运算符。

2．关系表达式

用关系运算符将两个操作数连接起来的表达式，称为关系表达式。例如，a>b, a!=b+c, 'a' > 'b' 等表达式都是合法的关系表达式。

关系表达式通常用于表达一个判断条件的结果，当条件成立时表达式的值为"真"，用数字"1"代表，当条件不成立时表达式的值为"假"，用数字"0"代表。

例如，设 a=5, b=4, c=3, 则：

（1）a<b。因为 a=5, b=4, a 小于 b 是不成立的，所以此表达式为假，值为 0。

（2）a!=b。因为 a=5, b=4, a 不等于 b 是成立的，所以此表达式为真，值为 1。

（3）a >b-c。因为数学运算符优先级高于关系运算符，因而先进行数学运算 b-c，计算结果为 1，比 a 的值要小，条件成立，所以此表达式为真，值为 1。

（4）a>b==c。因为 ">" 的优先级高于 "==" 的优先级，因而先进行 "a>b" 的关系运算，a 大于 b 是成立的，其结果为 1，再判断与 c 的值是否相等，c 不等

than that of the latter two. The priority of relational operators is lower than that of arithmetic operators, but higher than that of logic operators.

2. Relational Expression

The expression utilizing relational operators to connect two operation numbers is called relational expression. For example, a>b, a!=b+c, 'a'>'b' expressions are all legal relational expressions.

Relational expressions are generally applied for presenting one result for judging condition, the value of expression will be "true" when the conditions are available, utilizing the number "1" to present, the value of expression will be "false" when the conditions are unavailable, utilizing the number "0" to present.

For example, supposing a=5, b=4, c=3, then:

(1) a<b. Due to a=5, b=4, it is invalid for a is less than b, then the expression is false, the value is 0.

(2) a!=b. Due to a=5, b=4, it is valid that a is not equal to b, so this expression is true, the value is 1.

(3) a >b-c. Owing to the fact that the priority of mathematical operators is higher than that of relational operators, then the mathematical operation b-c will be carried out first, the calculation result is 1, which is less than the value of a, the conditions are valid, so the expression is true, the value is 1.

(4) a>b==c. Because the priority of ">" is higher than that of "==", then the relational operation of "a>b" will be carried out first, it is valid that a is greater than b, its result is 1, then judging if its value is equal

于1的，所以表达式为假，值为0。

注意：

在C语言中，用非0值表示"真"，用0值表示"假"。也就是说，不管表达式为何种类型，都可以作为条件判断，只要表达式的值为0，就表示表达式为假，即条件不成立；而只要表达式的值为非0，就表示表达式为真，即条件成立。

初学者经常会将条件表达式"a==b"写错成"a=b"，两个表达式虽然都可以作为条件判断，但其意义却是完全不同的。表达式"a==b"判断的是a与b是否相等，如果a、b相等则表达式为真；如果不相等则表达式为假。而表达式"a=b"是将变量b的值赋给变量a，然后用a的值来判断表达式的结果，如果变量a为非0值，则表达式为真；如果变量a为0，则表达式为假。

2.2.2 逻辑运算符与逻辑表达式

有时候判断的条件并不是一个简单的条件，而是由若干个简单的条件组合起来，组成一个复合条件。如，假设三条边长分别为a、b、c，判断其能否构成一个三角形。根据三角形构成的要求可以得知，构成三角形的三条边的要求是任意两条边的边长之和大于第三条边的边长，对此就需

to c, if c is not equal to 1, then the expression is false, the value is 0.

Attention:

In C language, utilizing the non0 value to present "true", applying the 0 value to present "false", that is, no matter the expression is of which type, it can be utilized as to conditions for judging, as long as the value of expression is 0, meaning the expression is false, namely, the conditions are invalid; While as long as the value of expression is not 0, signifying the expression is true, then the conditions are valid.

The beginners would always write the conditional expression "a==b" as "a=b" mistakenly, though these two expressions can both utilized as conditions for judging, their meanings are completely different. Applying the expression of "a==b" to judge if a is equal to b, if a is equal to b, then the expression is true; if a is not equal to b, then the expression is false. While the expression of "a=b" is to assign the value of variable b to variable a, then utilizing the value of a to judge the result of expression, if the variable a is not 0 value, then the expression is true; if the the variable a is 0, then the expression is false.

2.2.2 Logical Operator and Logical Expression

Occasionally the judging conditions don't belong to simple conditions, on the contrary, they are composed of several simple conditions, forming a compound condition. For example, supposing the length of three sides is a, b and c respectively, judging if it can form a triangular. In accordance with the composition conditions

Second Task: Grade Transformation of Student Score

要判断三个条件：a+b>c；b+c>a；c+a>b，只有这三个条件都满足,才能构成三角形。这个组合复合条件不能使用一个关系表达式来表示，需要使用逻辑运算符。

1. 逻辑运算符

在 C 语言中，有三种逻辑运算符，逻辑运算符见表 2.2。

for triangular, it can be known that the requirements for composing the three sides of triangular lie in the sum length of any two sides should be greater than that of the third side, in this regard, three conditions need to be judged: a+b>c, b+c>a, c+a>b, as long as these three conditions are met, thus forming a triangular. The compound conditions for this combination cannot be presented by one relational expression, the logical operator is required.

1. Logical Operator

In C language, there are three kinds of logical operators, the logical operators were presented in Table 2.2.

表 2.2 逻 辑 运 算 符

运算符	含义	举例	说明	优先级
!	逻辑非	!a	如果 a 为真，则 !a 为假；否则为真	高 ↓ 低
&&	逻辑与	a&&b	如果 a 和 b 都为真，则结果为真；否则为假	
||	逻辑或	a||b	如果 a 和 b 都为假，则结果为假；否则为真	

Table 2.2 Logical Operator

Operator	Meaning	Example	Description	Priority
!	Logical negation	!a	If a is true, then !a is false, or it is true	High ↓ Low
&&	Logical and	a&&b	If both a and b are true, then the result is true, or else it is false	
||	Logical or	a||b	If both a and b are false, then the result is false, or else it is true	

（1）运算数:运算符！是一元运算符，只需要一个操作数，运算符 && 和 ‖ 都是二元运算符，需要两个操作数。

(1) Operation Number: Operator ! It is unitary operator, only one operation number is required, operator && and ‖ are both binary operators, two operation numbers are required.

（2）结合性：运算符！是右结合，即从右往左运算，运算符&&和‖都是左结合，即从左往右运算。

（3）优先级：三个逻辑运算符的优先级都不相同，运算符！的优先级最高，运算符&&的优先级次之，运算符‖的优先级最低。

逻辑运算符、关系运算符、算术运算符的优先级顺序如图2.1所示。

(2) Associativity: operator! It is right-associative, namely, to operate from right to left, operator && and || are both left-associative, namely, to operate from left to right.

(3) Priority: the priority of three logical operators is not the same, the priority of operator! is the highest, the priority of operator && takes second place, the priority of operator || is the lowest.

The priority sequence of logical operator, relational operator and arithmetic operator was presented in Figure 2.1.

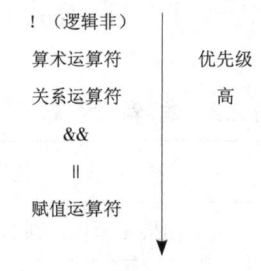

```
！（逻辑非）
算术运算符          优先级
关系运算符           高
   &&
    ‖
赋值运算符           ↓
```

图2.1 常用运算符的优先级顺序
Figure 2.1 The Priority Order of Commonly Utilized Operator

2. 逻辑表达式

用逻辑运算符将操作数连接起来的表达式，称为逻辑表达式。逻辑表达式常用于表达一个组合条件的判断结果，当条件成立时表达式的值为"真"，用数字"1"代表，当条件不成立时表达式的值为"假"，用数字"0"代表。逻辑运算符的真值表如表2.3所示。

2.Logical Expression

The expression applying logical operator to connect operation number is called logical expression. Logical expression is generally utilized to present the judgment result of a combination condition, when the conditions are valid, the value of expression is "true", utilizing the number "1" to present, when the conditions are invalid, the value of expression is "false", utilizing the number "0" to present. The truth table regarding logical operator was presented in Table 2.3.

Second Task: Grade Transformation of Student Score

表 2.3 逻辑运算符的真值表

a	b	!a	a&&b	a||b
非 0	非 0	0	1	1
非 0	0	0	0	1
0	非 0	1	0	1
0	0	1	0	0

Table 2.3 The Truth Table of Logical Operator

a	b	!a	a&&b	a||b
Non 0	Non 0	0	1	1
Non 0	0	0	0	1
0	Non 0	1	0	1
0	0	1	0	0

对于"逻辑与"运算，只有两个操作数都为真时，结果才为真，只要有一个操作数为假，结果就为假，在中文表达中用"同时""并且"进行描述，因此如果表示两个条件必须同时成立时,可用"逻辑与"运算符连接这两个条件。

As for "logical and" operation, only when the two operation numbers are true, the result can be true, as long as one operation number is false, the result will be false, utilizing "in the meantime" and "and" to describe in Chinese language expression, therefore, when signifying that the two conditions must be valid at the same time, the operator of "logical and" can be applied to connect these two conditions.

对于"逻辑或"运算，只要有一个操作数为真，结果就为真，只有两个操作数都为假时结果才为假,在中文表达中用"或者"进行描述，因此当表示两个条件可以进行二选一时，可以用"逻辑或"运算符连接这两个条件。

As for "logical or" operation, as long as one operation number is true, the result will be true, as long as both the two operation numbers are false, the result will be false, utilizing "or" to describe in Chinese language expression, therefore, when signifying that the two conditions can perform one-out-of-two arrangement, the operator of "logical or" can be applied to connect these two conditions.

例如：

若 x=15，y=20，则表达式 x&&y 的值为 1，因为 x、y 的值都是非 0，被认为是"真"。

若 a 是大于 0 且不大于 100 的数，则可用表达式"a>0&&a<=100"进行表示。

For example

If x=15, y=20, then the value of expression x&&y is 1, because the values of x and y are both non0, which is considered as "true".

If a is the number greater than 0 but less than 100, then the expression of "a>0&&a<=100" can be utilized to present.

若 a 和 b 至少有一个数是偶数，则可用表达式"a%2==0||b%2==0"进行表示。

若字符变量 c 是小写字母，则可用表达式"c>='a'&&c<='z'"进行表示。

2.2.3 条件运算符与条件表达式

在 C 语言中，根据判断一个条件是否成立而取不同的值，可采用条件运算符来实现。C 语言的条件运算符由"?"和":"组成，是 C 语言中唯一的一个三元运算符，是一种功能很强的运算符。采用条件运算符将两个不同运算分量连接起来的式子称为条件表达式。

条件表达式的格式为

表达式 1 ? 表达式 2 : 表达式 3

当表达式 1 的值为真时，整个表达式的值等于表达式 2 的值；当表达式 1 的值为假时，整个表达式的值等于表达式 3 的值。

例如 (x>=0)? 1 : -1，该表达式的值取决于 x 的值，如果 x 的值大于等于 0，该表达式的值为 1，否则表达式的值为 -1。

条件运算符的结合性是右结合，它的运算优先级低于算术运算符、关系运算符和逻辑运算符，高于赋值运算符。

If at least one number of a and b is even number, then the expression of "a%2==0||b%2==0" can be utilized to present.

If the character variable c is lowercase, then the expression of "c>='a'&&c<='z'" can be utilized to present.

2.2.3 Conditional Operator and Conditional Expression

In C language, obtaining different values by means of judging whether one condition is valid or not, it is available to apply conditional operation to realize it. The conditional operator of C language is composed of "?" and ":", which is the only ternary operator in C language and a operator with great function. The formula taking advantage of conditional operator to connect two different operational components is called conditional expression.

The formula of conditional expression is Expression 1? Expression 2: Expression 3

When the value of expression 1 is true, the value of whole expression is equal to the value of expression 2; When the value of expression 1 is false, the value of whole expression is equal to the value of expression 3.

Such as (x>=0)? 1 : -1, the value of this expression depends on the value of x, if the value of x is greater or equal to 0, then the value of this expression is 1, or else the value of expression is -1.

The conditional operator is right-associative, its operation priority is lower than arithmetic operator, relational operator and logical operator, and higher than assignment operator.

【例 2.1】使用条件运算符求两个数中的大值。

【解题思路】若要求出三个数中的最大值，可用表达式"a>b?a:b"作为求解 a、b 两个数中较大值的运算分量，再与第三个变量 m 比较大小，嵌套使用条件运算符，即可求解 a、b、m 三个变量中的最大值，此时表达式可用"m>(a>b?a: b)? m : (a>b?a:b)"进行表示。

【Example 2.1】Applying conditional operator to calculate the greater value of these two numbers.

【Problem-Solving Ideas】As for calculating the greatest value of these three numbers, the expression "a>b?a:b" can be utilized as the operation component for calculating the greater value of these two numbers of a and b, then comparing its value with the third variable m, the greatest value among these three variables of a, b and m can be calculated by means of nested application of conditional operator, at this point, the expression can be presented by "m >(a>b?a: b)? m: (a>b?a:b)".

```c
#include <stdio.h>
void main()
{
    int a=6,b=7,m;
    m=a>b?a:b;
    printf("%d、%d 两数中的大值为:%d\n",a,b,m);
}
```

2.2.4 单分支语句

当只有条件满足时才会进行下一步操作，此种情况为单分支结构，单分支选择结构采用简单的 if 语句实现，其格式为

```
if（条件表达式）
{
    语句组；
}
```

单分支结构的流程图如图 2.2 所示。if 语句的执行过程是先判断"条件表达式"是否成立，若条件成立，表示条件表达式

2.2.4 Simple Branched Statement

Only when the conditions are met the next operation can be carried out, it is simple branched structure in this case, simple branched selective structure can be realized by simple if statement, its format is:

```
if (conditional expression)
{
    Statement group;
}
```

The flow diagram of simple branched structure was presented in Figure 2.2. The operation procedure of if statement is to first judge if the "conditional expression"

的值为真，则执行语句组，否则什么也不做。语句组可以是简单的一条语句（此时大括号可以省略），也可以是多条语句构成的复合语句（此时大括号必须保留）。

is valid, if it is valid, the value presenting conditional expression is true, then performing statement group, or nothing needs to be done. Statement group can be a simple statement (at this point, the brace can be omitted), it can also be the compound statement composed of several statements (at this point, the brace must be preserved).

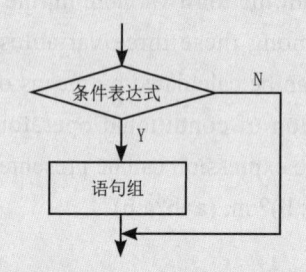

图 2.2 单分支结构的流程图

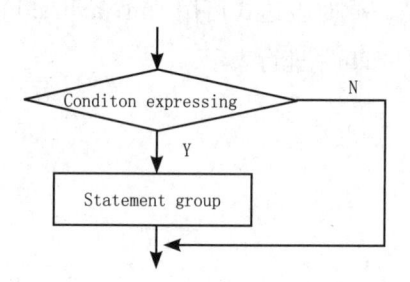

Figure 2.2　Flow Diagram of Simple Branched Structure

【例 2.2】从键盘输入一个数，输出其绝对值。

【解题思路】定义一个变量 x，用来接收用户从键盘输入的数据，并作为绝对值进行输出。从键盘输入数据后，判断 x 的值是否小于 0，如果 x 小于 0，则 y=-x，否则不做任何处理，最后输出 x。程序流程图如 2.3 所示。

2.2.5　双分支语句

不管条件是否满足，都会进行下一步操作，同时根据条件是否满足要进行不同的操作，此种情况为双分支结构，双分支选择结构采用 if-else 语句实现，其

【Example 2.2】Inputting one number from keyboard, outputting its absolute value.

【Problem-Solving Ideas】Defining one variable x to accept the data input by user from keyboard, then outputting it as the absolute value. After inputting date from keyboard, judging if the value of x is less than 0, if the x is less than 0, then y=-x, or else no treatment will be done, at last, outputting x. The flow diagram was presented in Figure 2.3.

2.2.5　Double-Branched Statement

No matter if the conditions are met, the next operation will be proceeded, in the meantime, different operations will be performed based on the fact that if the conditions are met, it is double-branched

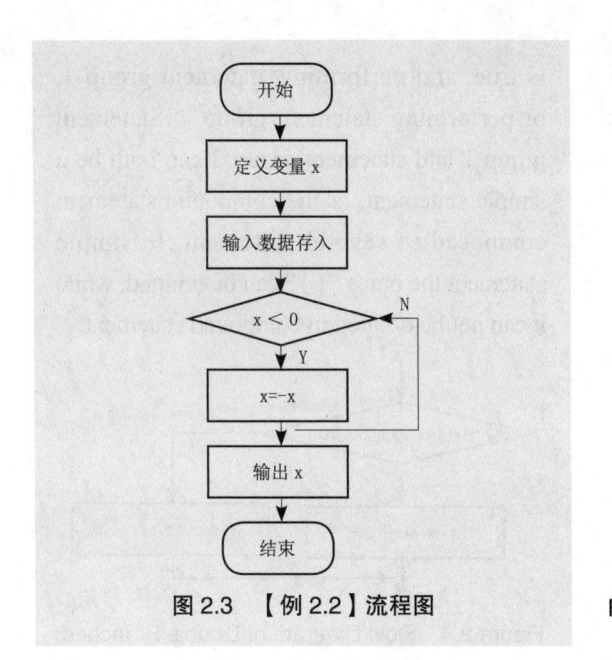

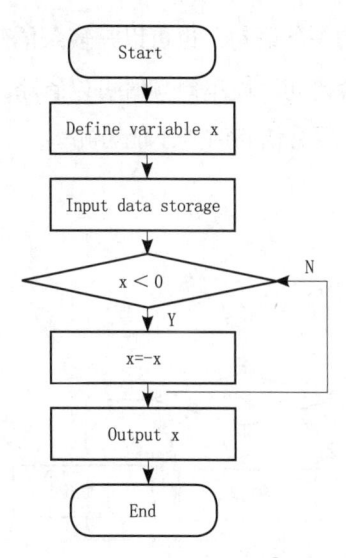

Second Task: Grade Transformation of Student Score

图 2.3 【例 2.2】流程图

Figure 2.3　The Flow Diagram of【Example 2.2】

```
1  #include <stdio.h>
2  void main()
3  {
4      int x;
5      printf("请输入一个整数：");
6      scanf("%d",&x);
7      if(x<0)
8          x=-x;
9      printf("其绝对值为：%d",x);
10 }
```

格式为

```
if（条件表达式）
{
    语句组 1；
}
else
{
    语句组 2；
}
```

双分支结构流程图如图 2.4 所示。if-else 语句的执行过程是先判断"条件表达式"是否成立，若条件成立，表示条件表达式的值为真，则执行语句组 1，否则执行语句组 2。语句组 1 和语句组 2 可以是

structure in this case, the double-branched selective structure is realized by if-else statement, its format is

```
if (conditional expression)
{
    Statement group 1;
}
else
{
    Statement group 2;
}
```

The flow diagram of double-branched structure was presented in Figure 2.4. The operation procedure of if-else statement is to first judge if the "conditional expression" is valid, if the conditions are valid, then the value for presenting conditional expression

简单的一条语句,也可以是多条语句构成的复合语句。简单语句时可以省略大括号"{ }",复合语句时则不能省略。

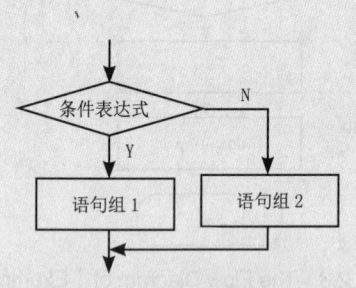

图 2.4 双分支结构流程图

【例 2.3】从键盘输入一个数,判断其奇偶性。

【解题思路】定义一个变量 x,用来接收用户从键盘输入的数据。从键盘输入数据后,判断 x%2 是否等于 0,如果等于 0,则 x 为偶数,否则 x 为奇数,最后输出 x 的奇偶性。程序流程图如 2.5 所示。

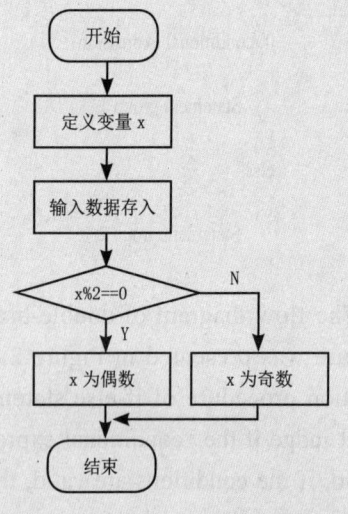

图 2.5 【例 2.3】的流程图

is true, and performing statement group 1, or performing statement group 2. Statement group 1 and statement group 2 can both be a simple statement, or the compound statement composed of several statement. In simple statement the brace "{ }" can be omitted, while it can not be omitted in compound statement.

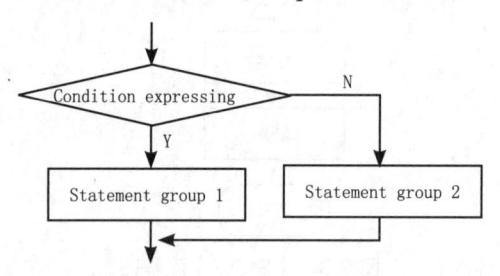

Figure 2.4 Flow Diagram of Double Branched Structure

【Example 2.3】 Inputting one number from keyboard to judge its parity.

【Problem-Solving Ideas】 Defining one variable x to accept the data input by user from keyboard. After inputting data from keyboard, judging if x%2 is equal to 0, if it is equal to 0, then x is even number, or else x is odd number, at last, outputting the parity of x. The procedure flow diagram was presented in Figure 2.5.

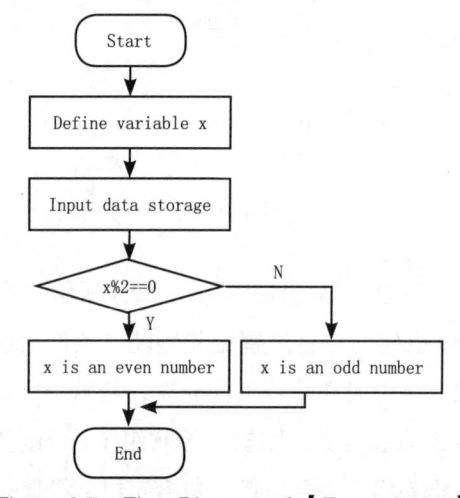

Figure 2.5 Flow Diagram of 【Example 2.3】

2.2.6 多分支语句

有些情况下，对于一个条件进行分析时，有多于两种的不同结果，针对每种结果有不同的操作，此种情况为多分支结构。多分支结构采用 if-else-if 语句或者 switch 语句实现。

1.if-else-if 语句

if-else-if 语句可以认为是 if-else 语句的 else 语句中又包含了一个 if-else 语句，相当于 if-else 语句的嵌套使用方法之一，其格式为

```
if（条件表达式 1）
{
    语句组 1;
}
else if（条件表达式 2）
{
    语句组 2;
}
……
else if（条件表达式 n）
{
    语句组 n;
}
else
{
    语句组 n+1;
}
```

此种形式多分支结构的流程图如图 2.6 所示。首先判断是否满足条件表达式 1，当条件表达式 1 的值为真时，执行语句组 1，否则，在条件表达式 1 的值为假的情况下再判断是否满足条件表达式 2，当条件表达式 2 的值为真时，执行语句组 2，

2.2.6 Multiple-Branched Statement

In some cases, while analyzing one condition, there will be two different kinds of results, aiming at the different operations of each result, it is multiple-branched structure in this case. if-else-if statement or switch statement can be realized by multiple-branched structure.

1.If-else-if Statement

if-else-if statement can be considered as another if-else statement included in else statement of if-else statement, which is equivalent to one of the embedding application methods for if-else statement, its format is:

```
if (conditional expression 1)
{
    Statement group 1;
}
else if (conditional expression 2)
{
    Statement group 2;
}
……
else if (conditional expression n)
{
    Statement group n;
}
else
{
    Statement group n+1;
}
```

The flow diagram for multiple-branched structure of this type was presented in Figure 2.6. First judging if the conditional expression 1 is satisfied, when the value of conditional expression 1 is true, performing statement group 1, on the contrary, in the case when the value of conditional expression 1 is false, then judging if the

否则，在条件表达式2的值为假的情况下再判断接下来的条件表达式是否满足，直到判断最后一个条件表达式 n 是否满足，当条件表达式 n 的值为真时，执行语句组 n，否则，执行语句组 n+1，此时意味着前面所有的条件表达式都不满足。

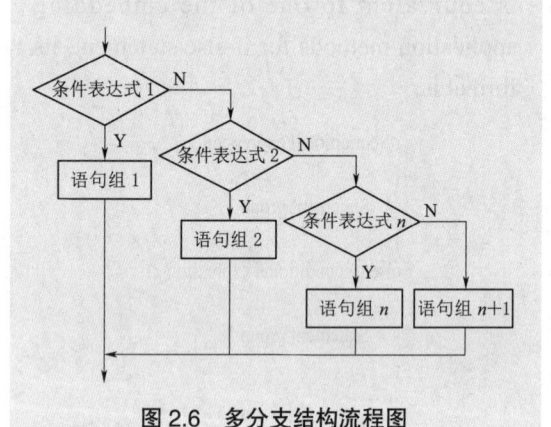

图2.6 多分支结构流程图

【例2.4】身体质量指数 BMI 简称体质指数，是国际上常用的衡量人体胖瘦程度以及是否健康的一个标准，其计算公式为：BMI=体重÷身高^2。（体重单位：千克；身高单位：米）BMI 正常值在20至25之间，20以下为偏瘦，超过25为超重，30以上则属肥胖。编程

conditional expression 2 is met, when the value of conditional expression 2 is true, performing statement group 2, or else, in the case when the value of conditional expression 2 is false, then judging if the following conditional expression is met, until judging if the last conditional expression *n* is satisfied, when the value of conditional expression *n* is true, performing statement group *n*, or else, performing statement group *n*+1, at this point, signifying that all of the former conditional expressions cannot be satisfied.

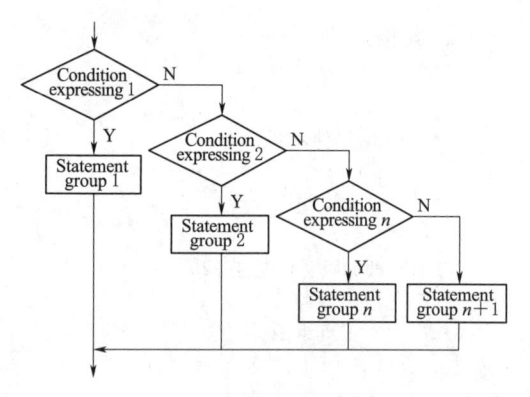

Figure 2.6 Multiple-Branched Structure Flow Diagram

【Example 2.4】Body mass index BMI is called body index for short, which is the commonly used standard for measuring the weight status of human body as well as the health condition, its calculation formula is: BMI=Weight÷Height^2 (Weight unit: kilogram; Height unit: meter) The normal value of BMI is between 20 and 25, signifying thinnish with points below 20, signifying overweight with points over 25, signifying fat with points over 30. Programming is realized by inputting

Second Task: Grade Transformation of Student Score

实现从键盘输入体重和身高，根据给定的公式计算 BMI 值，最后输出属于哪种体型。

【解题思路】定义三个单精度变量 w、h、BMI，其中 w 表示接收用户输入的体重，h 表示接收用户输入的身高，BMI 表示由公式计算得到的体质指数。从键盘输入数据后，根据公式 BMI=w/(h×h) 可以得到具体的体质指数 BMI 的值，若 BMI 的值小于 20，则输出此时的体质指数的值，并输出偏瘦体型；否则若 BMI 的值不小于 20 且小于等于 25，则输出正常体型；否则若 BMI 的值不小于 25 且小于等于 30，则输出超重体型；否则若 BMI 的值都不满足以上情况时，即 BMI 的值大于 30，则输出肥胖体型。

weight and height from keyboard, in light of the defined formula to calculate BMI value, at last, outputting the corresponding body type.

【Problem-Solving Idea】Defining three single precision variables w, h and BMI, among them, w signifies to accept the input weight by users, h signifies to accept the input height by user, BMI signifies the body index calculated by formula. After inputting data from keyboard, in light of the formula BMI=w/(h×h), the value of specific body index BMI can be obtained, if the value of BMI is less than 20, then outputting the value of body index at this point, and outputting the thinnish body form; or if the value of BMI is not less than 20 and is less than or equal to 25, then outputting the normal body type; or if the value of BMI is not less than 25 and is less than and equal to 30, then outputting the overweight body type; or if the value of BMI cannot satisfy the above condition, namely, the value of BMI is less than 30, then outputting fat body type.

```c
#include <stdio.h>
void main()
{
    float w,h,BMI;
    printf("请输入身高（米）和体重（千克）: ");
    scanf("%f%f",&h,&w);
    BMI=w/(h*h);
    if(BMI<20)
        printf("BMI=%.1f，偏瘦型。",BMI);
    else if(BMI<=25)
        printf("BMI=%.1f，正常型。",BMI);
    else if(BMI<=30)
        printf("BMI=%.1f，超重型。",BMI);
    else
        printf("BMI=%.1f，肥胖型。",BMI);
}
```

注意：

　　if-else 语句嵌套使用形式是非常灵活的，还可以在 if-else 语句的 if 语句中再包含一个或多个 if-else 语句，所以需要注重从语义的逻辑方面考虑如何组织 if 语句的嵌套，关键在于把握两个点：第一，不管是哪种形式的 if 语句，从语法的角度考虑，只是一条语句；第二，if、else 子句下的语句组可以是一般的语句，也可以是 if 语句。

　　另外，else 语句总是与它上面最近的、未曾配对的 if 语句配对，因此建议保留大括号，以增加程序的可读性，避免产生误解。

2.switch 语句

　　一般来说，当一个问题需要判断的情况大于三种时，通常使用 switch 语句实现多分支结构。switch 语句也称为开关语句，根据一个表达式可能产生的不同结果，选择其中一个或几个分支执行，switch 语句通常用于各种分类统计、菜单选择等程序设计。switch 语句的基本格式如下

```
switch（表达式）
{
    case 常量表达式 1:    语句组 1；
    case 常量表达式 2:    语句组 2；
    ……
```

Attention:

The nesting form of if-else statements is quite flexible, and one or more if-else statements can be included in the if statements of if-else statements. Therefore, it is necessary to pay attention to how to organize the nesting of if statements from the perspective of semantic logic. The key lies in grasping two points: first, no matter what form of if statements, they are only one statement from the perspective of grammar; second, the statement groups under the if and else clauses can be general statements or if statements.

In addition, the else statement is always paired with the nearest unpaired if statement on it, so it is suggested to keep braces to increase the readability of the program and avoid misunderstanding.

2.Switch Statement

Generally speaking, when the conditions need to be judged for one problem are over three kinds, the multiple-branched structure will be realized by switch statement generally. Switch statement is also called switch caset, since one expression can produce different results, selecting one or several branches from them to perform, switch statement can be generally utilized for program designs such as various classified statistic and menu selection. The basic format of switch statement is as following

```
switch (expression)
{
    case constant quantity expression 1:
                    Statement group 1;
    case constant quantity expression 2:
                    Statement group 2;
    ……
```

```
        case constant quantity expression n:
                        Statement group n;
        default:        Statement group n+1;
}
```

While the program is performing switch statement, it will first calculate the value of "expression" within the brackets followed by switch, then looking for the constant quantity expression followed by case equal to this value from top to bottom, when the value of "expression" is equivalent to the value of constant quantity expression followed by certain case, then starting from this case statement to perform all the latter statements from top to bottom, including all of the latter case statements, no longer performing the judgement for the value of "expression" and constant quantity expression, until encountering break statement or the switch statement ends. Therefore, when each case statement is mutually independent and exclusive, each case statement will be followed by break statement to pop out switch structure compulsively. If there is no case constant quantity equal to switch statement, the program will then go to perform the statements followed by default mark, no default branch is indispensable in switch structure.

【Example 2.5】Inputting number from 1 to 7 on the keyboard, outputting the English letter of week number represented by this number, for example, number 1 represents Monday, number 7 represents Sunday.

【Problem-Solving Ideas】There are seven branch numbers for this problem, then applying switch structure to realize it. As for defining a variable to accept the number input by user on keyboard, taking this

个 case 分支后不同的 1~7 的数字，代表着不同的星期几，每个分支都是互斥的，因此每个 case 分支后都要加上 break 语句。

variable as conditional expression to judge, the different 1 to 7 numbers behind each case branch present different week numbers, each branch is mutually exclusive, therefore, adding break statement behind each case branch.

```c
#include <stdio.h>
void main()
{
    int i;
    printf("请输入数字（1~7）：");
    scanf("%d",&i);
    switch(i)
    {
        case 1:printf("Monday");break;
        case 2:printf("Tuesday");break;
        case 3:printf("Wednesday");break;
        case 4:printf("Thursday");break;
        case 5:printf("Friday");break;
        case 6:printf("Saturday");break;
        case 7:printf("Sunday");break;
        default:printf("输入数字有误！");
    }
}
```

注意：

switch 后面括号内的"表达式"的值要求为整数，因此该表达式一般是整型或字符型，而且每个 case 后面的"常量表达式"的类型应该与 switch 表达式值的类型一致。

每个 case 的常量表达式的值是对 switch 表达式的枚举，必须互不相同。若每个 case 分支的最后都加上了 break 语句，则 case 分支的排列次序可以任意安排，不会改变程序执行的结果。若 case 分支的最后没有 break 语句，则 case 分支的排列次序可能会影响程序执行的结果。

switch 语句常用于菜单选择的程序设计，配合函数的使用，可用于设计多任务

Attention:

The value of "expression" within the brackets following switch must be integer, therefore, such expression is generally of integer type or character type, and the type of "constant quantity expression" following each case should be consistent with the type of switch expression value.

The value of constant quantity expression for each case is the enumeration for switch expression, and it must be different. If break statement is attached to the last part of each case branch, then the ordering of case branch can be arranged randomly, which can not change the result for performing program. If no break statement can be found in the last part of case branch, then the ordering of case branch may affect the result of program performing probably.

switch statement is generally applied for program designing of menu selection,

Second Task: Grade Transformation of Student Score

的学生成绩管理系统等。

【例2.6】编写一个菜单显示程序，界面如下。要求输入1~4时进行相应的显示，如输入1则显示"你选择了1.添加记录"，输入2则显示"你选择了2.删除记录"，当输入0~4以外的数据时，显示"选择错误！"。

combining with the application of function, which can be utilized for designing the student score management system with multiple tasks.

【Example 2.6】Compiling one menu display program, the interface is as following:

It is required to input 1 to 4 for corresponding display, if 1 is input, then it will display "you choose 1, adding record", inputting 2 to display "you choose 2, deleting record," when inputting data beyond 0 to 4, then displaying "selection error!".

```
        主菜单
1. 添加记录    2. 删除记录
3. 查找记录    4. 预览记录
请选择 1～4，按 0 退出：
```

【解题思路】定义一个变量i，用来接收用户从键盘输入的选择。有0~4五个合法的输入选择，用switch结构实现，每个case的常量分别为0~4，因为每个分支都是互斥的，因此每个case分支后加break语句结束switch语句，除此之外的数据输入对应着default语句。

【Problem-Solving Ideas】Defining one variable i to accept the selection input by user from keyboard. There are five legitimate input selections from 0 to 4, which can be realized by switch structure, the constant quantity of each case is from 0 to 4, because each branch is mutually exclusive, hence, break statement is added to each case branch for ending switch statement, apart from this, the data input is corresponded to default statement.

```c
#include <stdio.h>
void main()
{
    int i;
    printf("        主菜单\n");
    printf("1.添加记录   2.删除记录\n");
    printf("3.查找记录   4.预览记录\n");
    printf("请选择1~4，按0退出: ");
    scanf("%d",&i);
    switch(i)
    {
        case 0: printf("Bye! ");break;
        case 1: printf("你选择了1.添加记录\n");break;
        case 2: printf("你选择了2.删除记录\n");break;
        case 3: printf("你选择了3.查找记录\n");break;
        case 4: printf("你选择了4.预览记录\n");break;
        default:printf("选择错误\n");
    }
}
```

2.3 任务实现

【任务要求】某班级进行了一次 C 语言程序设计考试，教师按百分制给出了学生成绩，现在学校要求改为五级制进行打分，即 90~100 分为 A, 80~89 分为 B, 70~79 分为 C, 60~69 分为 D, 60 分以下为 E。分数可以任意输入。

【任务分析】考试分数为 0~100, 不在此范围内的分数应当被视为输入不合理，因此在进行分数等级转换之前要进行输入成绩合法性的判断。将百分制分数改为五级制，是根据 10 分一个阶段进行分级，因此将输入的有效分数分离出十位数字和个位数字，当十位数字为 10 或 9 时为"A"等级，当十位数字为 8 时为"B"等级，当十位数字为 7 时为"C"等级，当十位数字为 6 时为"D"等级，当十位数字为 0~5 时为"E"等级。

2.3.1 输入学生成绩，判断其合法性

只有当输入的成绩大于等于 0 并且小于等于 100 时，成绩才是合法的，所以需要用"逻辑与"将两个关系表达式

2.3 Task Implementation

【Task Requirement】One C Language Program Design examination has been carried out in certain class, the teachers rated student scores based on hundred-mark system, currently, it is required by school to change into five-class system, namely, obtaining 90 to 100 points is equivalent to A, obtaining 80 to 89 points is equivalent to B, obtaining 70 to 79 points is equivalent to C, obtaining 60 to 69 points is equivalent to D, obtaining points below 60 is equivalent to E. The points can be input randomly.

【Task Analysis】The examination score is between 1 to 100, it should be considered as unreasonable input if the scores are not within this range, therefore, before carrying out score grade conversion, it is necessary to perform the judgement for input grade legitimacy. Changing the hundred-mark system into five-class system, it is to rate based on the rule of 10 points for one grade, therefore, separating the effective input scores from tens digit numbers and units digit numbers, when the tens digit number is 10 or 9, then it belongs to "A" grade, when the tens digit number is 8, it belongs to "B" grade, when the tens digit number is 7, it belongs to "C" grade, when the tens digit number is 6, it belongs to "D" grade, when the tens digit number is between 0 to 5, it belongs to "E" grade.

2.3.1 Inputting Student Score, Judging its Legitimacy

Only when the input score is greater than and equal to 0 as well as less than and equal to 100, the scores can be legitimate, so it is necessary to utilize "logic and" to connect

score>=0 和 score<=100 连接起来，即可实现成绩合法的条件判断，否则代表着输入的成绩为不合法，可以用 if-else 语句实现。

these two relational expressions score>=0 and score<=100, then realizing the legitimate conditional judgment of scores, or else it will present that the input scores are not legal, which can be realized by if-else statement.

```
1   #include <stdio.h>
2   void main()
3   {
4       float score;
5       printf("请输入成绩：");
6       scanf("%f",&score);
7       if(score>=0&&score<=100)
8           printf("成绩合法。");
9       else
10          printf("成绩不合法。");
11  }
```

2.3.2 学生成绩等级转换

在合法成绩下将百分制转换成五级制，一共有 5 个分支，因此采用 switch 结构来实现。输入的成绩为 float 型，需要分离出十位的整数数字，因此先将 float 型的分数强制转换成 int 型后，再进行对 10 做整除运算。当对 10 做整除运算后的结果为 10 和 9 时，都是 A 等级，因此 case 后常量为 10 和 9 的分支具有相同的输出语句，从优化程序的角度出发，可以只保留 case 9 分支的语句，加上 break 语句结束 switch 结构。当对 10 做整除运算后的结果分别为 8、7、6 时，分别输出不同的等级，因此每条 case 分支后需要加上 break 语句结束 switch 结构。剩下的合法分数均为 E 等级，可以安排在 default 分支。

2.3.2 Grade Conversion of Student Scores

Changing hundred-mark system into five-class system under legal scores, there are mainly five branches, therefore, it can be realized by switch structure. The input scores belong to float type, it is necessary to separate the integer number of tens digit, therefore, first transforming the scores of float type into int type compulsively, after that, performing the exact division on 10. When the results obtained after exact division on 10 are 10 and 9, they belong to A grade, therefore, the branches with constant quantities of 10 and 9 following case have the same output statements, starting from the perspective of optimizer, it is available to only preserve the statements of case 9 branch, adding break statement to end switch structure. After exact division on 10, the results obtained are 8, 7 and 6 respectively, outputting different grades, therefore, break statement should be added to each case branch for ending switch structure. The left legal scores belong to E grade, which can be arranged on default branch.

任务2 学生成绩等级转换

```c
#include <stdio.h>
void main()
{
    float score;
    int i;
    printf("请输入成绩：");
    scanf("%f",&score);
    if(score>=0&&score<=100)
    {
        i=(int)score/10;
        switch(i)
        {
            case 10:
            case 9: printf("A等级\n");break;
            case 8: printf("B等级\n");break;
            case 7: printf("C等级\n");break;
            case 6: printf("D等级\n");break;
            default:printf("E等级\n");
        }
    }
    else
        printf("成绩不合法。");
}
```

【练习与提高】

1. 编写一个程序，输入 x 的值，按下列分段函数计算并输出 y 的值。

$$y = \begin{cases} x & x \leqslant 1 \\ 2x-1 & 1 < x < 10 \\ 3x-11 & 10 \leqslant x \end{cases}$$

2. 从键盘输入坐标图中一个点的坐标，判断它属于哪个象限。

3. 从键盘输入三角形三条边的大小，判断其是否能组成一个三角形，若可以，则输出此三角形的面积和类型（等边、等腰、直角、一般三角形）。[提示：假设三角形的三条边长分别为 a、b、c，p 为周长的一半，三角形面积可用公式

【Practice and Improvement】

1. Compiling one program, inputting the value of x, calculating and outputting the value of y according to the following piecewise function.

$$y = \begin{cases} x & x \leqslant 1 \\ 2x-1 & 1 < x < 10 \\ 3x-11 & 10 \leqslant x \end{cases}$$

2. Inputting the coordinate of one point on the coordinate graphs from keyboard, judging which quadrant it belongs to.

3. Inputting the length of three sides for triangular from keyboard, judging if it can form a triangular, if it can be realized, then outputting the area and type of this triangular (equilateral triangle, isosceles triangle, right triangle and ordinary triangular.) (Hint: Supposing that the length of three sides for triangular is a, b and c respectively, p is the half of perimeter, the area of triangular can be obtained by the formula

Second Task: Grade Transformation of Student Score

$s=\sqrt{p(p-a)(p-b)(p-c)}$ 得出。C 语言中求开根号，用 sqrt( ) 函数实现，j=sqrt(i) 表示对变量 i 开根号后的值赋给变量 j。]

4. 请编程实现从键盘输入某位员工的全月总收入额，按公式计算并显示员工的实发工资、应缴的个人所得税。目前我国公民个人工资、薪金所得税的计算式是根据个人月收入分级超额进行的，共分为 7 级，月收入不超过 5000 元的免交个人所得税，全月应纳税所得额为总收入减去 3500，个人所得税 = 全月应纳税所得额 * 适用税率 – 速算扣除数。

of $s=\sqrt{p(p-a)(p-b)(p-c)}$. Obtaining the square root calculations in C language, which can be realized by sqrt( ) function, j=sqrt(i) signifies to assign the value after square root calculations of variable i to variable j.)

4. Please realize to input the whole month total income of certain personnel from keyboard by programming, calculating and presenting the actual paid salary and the income tax payable of personnel according to the formula. At present, the calculation formula for personal salary and income tax payable of citizen in our country is performed by excessive monthly income rating, which can be classified into seven grades, the people with monthly income less than 5000 yuan doesn't need to pay the income tax payable, the whole month income tax payable should be the total income deducts 5000 yuan, income tax payable=whole month income tax payable*applicable tax rate-quick deduction.

级数	应纳税所得额（含税）
1	不超过 3000 元
2	超过 3000 元至 12000 元的部分
3	超过 12000 元至 25000 元的部分
4	超过 25000 元至 35000 元的部分
5	超过 35000 元至 55000 元的部分
6	超过 55000 元至 80000 元的部分
7	超过 80000 元的部分

Grade	Income Tax Payable (Containing Tax)
1	No more than 3000 yuan
2	The part between 3000 yuan and 12000 yuan
3	The part between 12000 yuan and 25000 yuan
4	The part between 25000 yuan and 35000 yuan
5	The part between 35000 yuan and 55000 yuan
6	The part between 55000 yuan and 80000 yuan
7	The part exceeds 80000 yuan

税率 /%	速算扣除数
3	0
10	210
20	1410
25	2660
30	4410
35	7160
45	15160

Tax Rate /%	Quick Deduction
3	0
10	210
20	1410
25	2660
30	4410
35	7160
45	15160

5. 水果店售卖各种品质的苹果，一级苹果 16.8 元 /kg，二级苹果 12.8 元 /kg，三级苹果 9.8 元 /kg，四级苹果 3.8 元 /kg。顾客购买某一级别的苹果后买单，收银员需要从键盘输入苹果的等级、重量、顾客的付款额，编程实现在屏幕上显示苹果的等级、重量、顾客应付款额、找零的钱数等内容。要求程序能正确处理任何数据，当输入苹果的等级、数量、顾客的付款额不合要求时，在屏幕上显示"数据输入错误"，程序结束。

6. 输入一个日期的年月日，计算并输出这天是该年的第几天。

5. Apples with different qualities are sold in fruit store, the first-level apple is 16.8 yuan/kg, the second-level apple is 12.8 yuan/kg, the third-level apple is 9.8 yuan/kg, and the fourth-level apple is 3.8 yuan/kg. The customers would pay the money after buying apples of each level, the cashier needs to input the grade and weight of apples as well as the payment amount of customers from the keyboard, the programming can display the grade and weight of apples, the payment amount of customers as well as the change on the screen. It is required that the program can deal with any data correctly, when the input grade and weight of apples as well as the payment amount of customers are not desirable, it will display "data input error" on the screen, and the program will end.

6. Inputting the full date, calculating and outputting it belongs to which day of this year.

任务 3
学生成绩分组汇总

Third Task: The Grouping and Summarizing of Student Scores

【知识目标】
1. 熟知循环的三种语句及其执行流程。
2. 掌握循环结构程序设计的基本方法。

【能力目标】
1. 能运用三种循环语句实现循环结构的程序设计。
2. 掌握三种循环之间的差异。

【重点、难点】
1. for 循环的使用。
2. 三种循环的灵活应用。

【课程思政】
1. 通过三种循环的程序设计，培养严谨的逻辑思维分析能力和应用能力，同时加强学生树立踏实、严谨细致的工作作风。
2. 通过递归函数的定义，强调言传身教的重要性。

【推荐教学方法】
通过教学做一体化教学，结合生活中常见的事例，使学生掌握知识点，学会编制程序流程图并进行程序的编写。

【推荐学习方法】
通过完成任务，在做中学、学中做，掌握实际技能与相关知识点。

【Knowledge Objective】
1. Having a good command of the three kinds of statements and their implementation procedure for loop.
2. Proficient in the basic methods of loop structure program design.

【Competency Objective】
1. Being able to apply the three kinds of loop statements to realize the program design of loop structure.
2. Proficient in the difference among three kinds of loops.

【Focal and Difficult Points】
1. The application of for loop.
2. The flexible application of three kinds of loop.

【Curriculum Ideological and Political Education】
1. By means of the program design of three kinds of loop, cultivating the cautious logical thinking analysis capability as well as the application capability, in the meantime, cultivating the steadfast and cautious working style of students.
2. By means of the recursion for the definition of function, emphasizing the importance of teaching by personal example as well as verbal instruction.

【Recommended Teaching Method】
By means of integrated teaching, combining the common cases in daily life, making the students grasp the knowledge points, learning to compile program flow diagram and edit the program.

【Recommended Learning Method】
By means of accomplishing tasks, learning by doing, doing by learning, mastering the actual technologies and the relevant knowledge points.

3.1 任务描述

某班中有四个小组，求本学期期中考试中每个小组数学成绩的总分及平均分。假设每个小组的成员数一样，都是8位，程序的运行要求输入一组同学的成绩后输出学生组员的平均分与总成绩。

针对上述需求进行分析，显然要求完成一个班中每个小组的学生数学成绩的平均分与总分的计算，首先必须做到在一个小组中对学生数学成绩的平均分与总分进行计算，然后重复4次。

3.1 Task Description

There are four groups in certain class, calculating the total and average scores for mathematics of each group in this mid-term examination. Supposing that the number of group member is the same eight for four groups, according to the operation requirements of program, after inputting the scores of students in one group, outputting the average and total scores of its group members.

Aiming at the above requirements to conduct analysis, it is necessary to accomplish the calculation of average and total scores for mathematics of students in each group in the class, first of all, it is necessary to accomplish the calculation average and total scores for mathematics of students in one group, then repeating for four times.

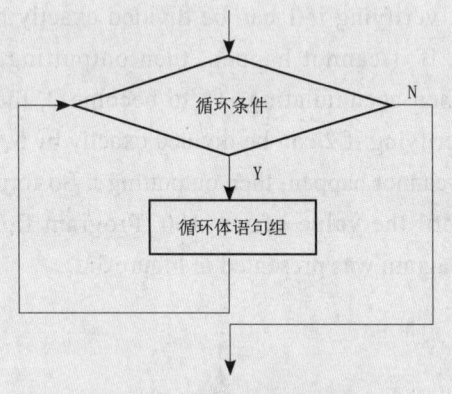

3.2 Relevant Knowledge

3.2.1 while Loop

1. General Format

```
while(loop condition){
    Loop body statement group
}
```

2. Implementation Procedure

(1) Solving "loop condition", if the solved result is logical true, then turning to (2); or turning to (3).

(2) Performing loop body statement group, then turning to (1).

(3) Exiting and ending loop.

The loop flow diagram was presented in Figure 3.1.

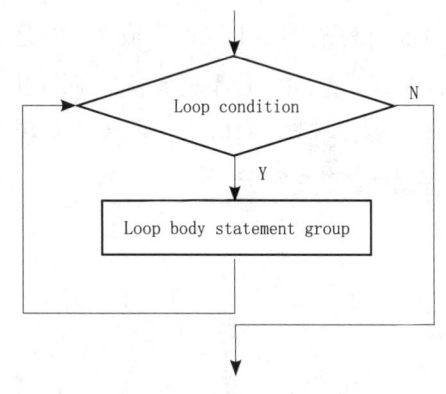

Figure 3.1　while Loop Flow Diagram

Analyzing the implementation procedure of the following program:

```
int m=1;
int sum = 0;
while(m<=10){
    scanf("%d",&score);
    sum += score;
    m++;
}
```

任务3 学生成绩分组汇总

首先，m 的初始值为 1，显然 1<=10，满足循环条件，执行循环语句组，即输入 score，然后将 score 累加到 sum 中，然后 m 的值加 1 为成了 2，继续判断是否满足循环条件，显然 2<=10，将继续执行循环语句组，直到 m=11，不再满足循环条件时为止。所以这段程序是输入 10 次 score，并将它们累加求总分。

【例 3.1】将 1 至 100 之间不能被 3 整除的数输出。

【解题思路】定义变量 i 并赋初始值为 1，验证 1 是否能被 3 整除，若不能整除，则输出 i；然后将 i+1 累加变成 2，再验证 2 否能被 3 整除，若不能整除，则输出 i；依此类推，直到 i 的值 <=100 为止，程序流程图如图 3.2 所示。

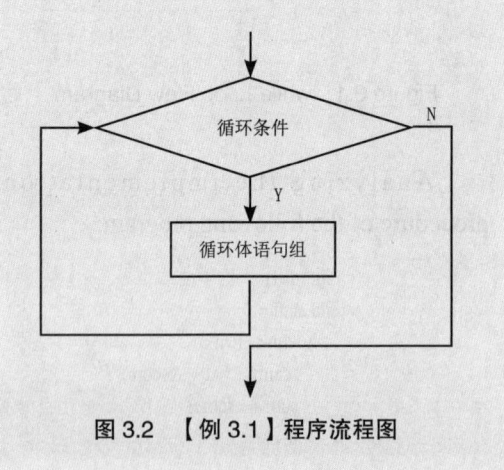

First of all, the initial value of m is 1, it is clear that 1<=10, meeting the loop condition, performing loop statement group, that is to input score, then accumulating score to sum, then adding the value 1 of m to become 2, continuing to judge if it meets the loop condition, it is clear that 2<=10, continuing to perform loop statement group, until m=11, when the loop conditions can not be satisfied. As for this program, it is to input score for ten times, and accumulating them to calculate the total scores.

【Example 3.1】Outputting the number between 1 to 100, which can not be divided exactly by 3.

【Problem-Solving Ideas】Defining variable i and assigning the initial value 1, verifying if 1 can be divided exactly by 3, if it cannot happen, then outputting i; Then accumulating i+1 to become 2, then verifying if 2 can be divided exactly by 3, if it cannot happen, then outputting i; So forth, until the value of i <=100. Program flow diagram was presented in Figure 3.2.

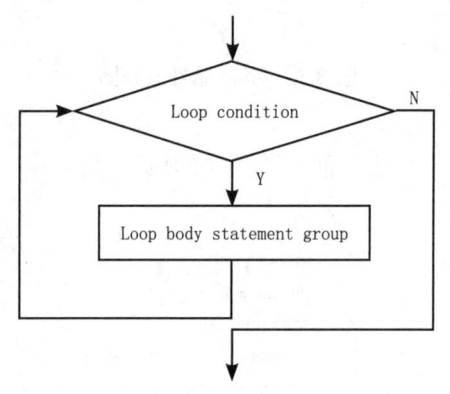

图 3.2 【例 3.1】程序流程图

Figure 3.2 【Example 3.1】Program Flow Diagram

50

Third Task: The Grouping and Summarizing of Student Scores

参考程序如下：

The referential program is as following:

```
1  #include <stdio.h>
2  main()
3  {
4      int i=1
5      while(i<=100)
6      {
7          if(i%3 != 0) printf("%3d",i);
8          i++;
9      }
10 }
```

【例3.2】求 sum=1+2+3+…+100。

【解题思路】求 sum=1+2+3+…+100，可以分解为

   sum=0;
   sum=sum+1;
   sum=sum+2;
   …
   sum=sum+100;

所以，可以看成 sum=sum+i 重复了100次，而 i 的值从1递增到100，为了使思路更清晰，可以画出本题具体实现流程图，如图3.3所示。

【Example 3.2】Calculating sum=1+2+3+…+100.

【Problem-Solving Ideas】Calculating sum=1+2+3+…+100, which can be decomposed as:

   sum=0;
   sum=sum+1;
   sum=sum+2;
   …
   sum=sum+100;

Therefore, it can be regarded as the sum=sum+i is repeated for one hundred times, while the value of i increases from 1 to 100, for making the idea clearer, the specific implementation flow diagram of this problem can be drawn, as it was presented in Figure 3.3.

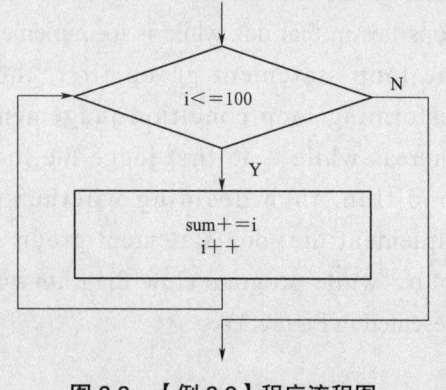

图3.3　【例3.2】程序流程图

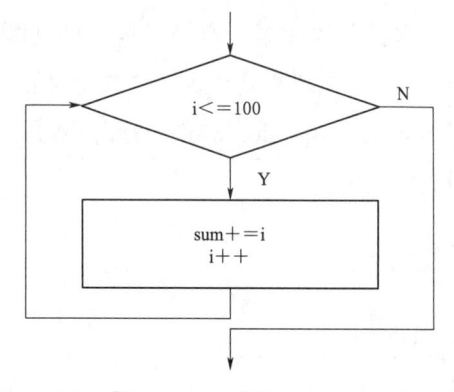

Figure 3.3　【Example 3.2】Program Flow Diagram

参考程序如下：

```c
#include <stdio.h>
main()
{
    int i=1,sum=0;
    while(i<=100)
    {
        sum += i;
        i++;
    }
    printf("sum=1+2+3+… +100的和为：%d\n",sum);
}
```

The referential program is as following:

3.2.2 do...while 循环

1. 一般格式

do{
　　循环体语句句组
}while(循环条件)

2. 执行过程

（1）执行循环体语句句组。

（2）求解 while 后面的"循环条件"，如果求解结果为逻辑真，则转到（1）；否则转到（3）。

（3）退出结束循环。

由上述执行过程可知，do...while 与 while 循环明显的区别是 do...while 是先执行一次循环语句句组，再做循环条件判断，而 while 是先判断循环条件再决定是否执行循环语句句组。do...while 循环流程图如图 3.4 所示。

3.2.2 do...while Loop

1. General Format

do{
　　Loop body statement group
}while(loop condition)

2. Implementation Procedure

(1) Implementing loop body statement group.

(2) Calculating the "loop conditions" following while, if the calculation result is logical true, then turning to (1); or else turning to (3).

(3) Exiting and ending loop.

It can be observed from the above implementation procedure that the clear difference between do...while and while loops lies in that do...while is to implement one loop statement group first, then performing loop condition judgement, whereas while is to first judge the loop condition, then deciding whether to implement the loop statement group or not.o...while program flow diagram was presented in Figure 3.4.

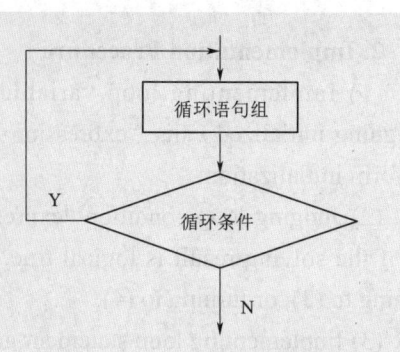

Third Task: The Grouping and Summarizing of Student Scores

图 3.4 do...while 循环流程图

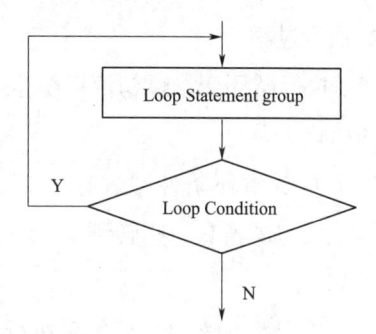

Figure 3.4 do...while Loop Flow Diagram

前面的【例 3.1】用 do...while 循环可改写为：

The above【Example 3.1】can be rewritten by do...while as:

```
#include <stdio.h>
main()
{
    int i=1;
    do{
        if(i%3 != 0) printf("%3d",i);
        i++;
    }while(i<=100);
}
```

前面的【例 3.2】用 do...while 循环可改写为：

The above【Example 3.2】can be rewritten by do...while as:

```
#include <stdio.h>
main()
{
    int i=1,sum=0;
    do
    {
        sum += i;
        i++;
    }while(i<=100);
    printf("sum=1+2+3+… +100的和为: %d\n",sum);
}
```

3.2.3 for 循环

1. 一般格式

for(表达式 1；表达式 2；表达式 3){
    循环体语句组
}

3.2.3 for Loop

1. General Format

for(Expression 1; Expression 2; Expression 3){
    Loop body statement group
}

2. 执行过程

（1）执行循环变量赋初值"表达式1"进行初始化。

（2）判断循环条件"表达式2"，如果求解结果为逻辑真，则转到（3）；否则转到（4）。

（3）执行循环语句组，然后执行循环变量增量"表达式3"后转（2）。

（4）退出结束循环。

while 循环流程图，如图3.5所示。

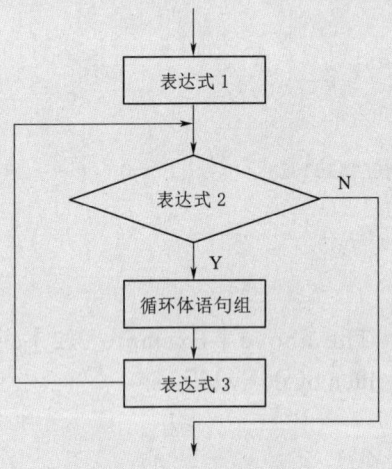

图 3.5 while 循环流程图

前面的【例3.1】用for循环可改写为：

2. Implementation Procedure

(1) Implementing loop variable and assigning initialized value "expression 1" to perform initialization.

(2) Judging loop condition "expression 2", if the solving result is logical true, then turning to (3); or turning to (4).

(3) Implementing loop statement group, then performing loop variable increment "expression 3" and turning to (2).

(4) Exiting and ending loop.

The program flow diagram of for statement was presented in Figure 3.5.

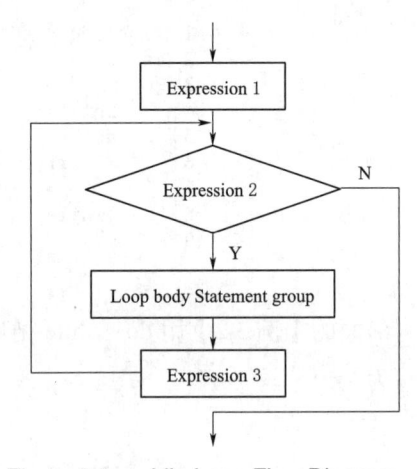

Figure 3.5 while Loop Flow Diagram

The former 【Example 3.1】 can be written by for loop as:

```c
#include <stdio.h>
main()
{
    int i;
    for(i=1;i<=100;i++)
    {
        if(i%3 != 0) printf("%3d",i);
    }
}
```

前面的【例3.2】用for循环可改写为：

```
1  #include <stdio.h>
2  main()
3  {
4      int i,sum=0;
5      for(i=1;i<=100;i++)
6      {
7          sum +=i;
8      }
9      printf("sum=1+2+3+… +100的和为：%d\n",sum);
10 }
```

使用for语句时，需要注意以下问题：

（1）for循环中的表达式1（循环变量赋初值）、表达式2（循环条件）、表达式3（循环变量增量）都是选择项，可以省略，但表达式分隔符号";"不能省略。

（2）省略表达式1（循环变量赋初值），表示不对循环控制变量赋初值。

（3）省略表达式2（循环条件），则程序将可能进入死循环。例如：

```
for (i=1;;i++)sum+=i;
```

相当于：

```
while(1)
{
    sum += i;
    i++;
}
```

（4）省略表达式3（循环变量增量），则不对循环控制变量进行修改，这时可在循环体语句组中加入修改循环控制变量的语句。例如：

The former 【Example 3.2】 can be written by for loop as:

While utilizing for statement, the following problems need to be laid emphasis on:

(1) The expression 1 in for loop (assigning initialized value to loop variable), expression 2 (loop condition), expression 3 (loop variable increment) are all selective options, which can be omitted, but the expression separator ";" can not be omitted.

(2) Omitting expression 1 (assigning initialized value to loop variable), signifying that no assigned initialized value has been imposed on loop control variables.

(3) Omitting expression 2 (loop condition), then the program would enter endless loop. For example

```
for(i=1;;i++) sum += i;
```

Equivalent to

```
while(1)
{
    sum += i;
    i++;
}
```

(4) Omitting expression 3 (loop variable increment), then don't make modification on loop control variables, at this point, the statement can be added to loop body statement group for modifying loop control variables. For example

```
for(i=1;i<=100;){sum+=i;i++;}
```

（5）省略表达式1（循环变量赋初值）和表态式3（循环变量增量）。例如：

```
for(;i<=100;) {
    sum += i;
    i++;
}
```

相当于：

```
while(i<=100){
    sum += i;
    i++;
}
```

3.2.4 循环嵌套

循环的嵌套又称为多重循环，就是在一个循环体内包含另外一个循环体。3种循环结构（while 循环、do…while 循环和 for 循环），不仅可以实现自身循环嵌套，还可以相互嵌套。一个循环体内包含另一个完整的循环体，称为嵌套循环或多层循环。在多重循环中内层的优先级比外层高，只有内层的循环执行完才能执行外层的，循环嵌套的要领对各种语言都适用。

几种常见的循环嵌套形式如下：

形式1：

```
while()
{
    while()
    {
        ...
    }
}
```

```
for(i=1;i<=100;){sum+=i;i++;}
```

(5) Omitting expression 1 (assigning initialized value to loop variable) and expression 3 (loop variable increment). For example:

```
for(;i<=100;) {
    sum += i;
    i++;
}
```

Equivalent to:

```
while(i<=100){
    sum += i;
    i++;
}
```

3.2.4 Loop Embedding

Loop embedding is also called multiple loop, which means to include another loop body in certain loop body. The three kinds of loop structures (while loop, do…while loop and for loop), they can not only realize their own loop embedding, but also performing embedding mutually. To include another complete loop body in certain loop body, it is called embedding loop or multiple loop. The priority of internal layer in multiple loop is higher than that of external layer, only when the internal loop has been performed, the external loop can be carried out, the technology for loop embedding is applicable to various languages.

Next, let's have a look at several common loop embedding forms.

Form 1：

```
while()
{
    while()
    {
        ...
    }
}
```

形式 2：

```
for(;;)
{
    for(;;)
    {
    ...
    }
}
```

形式 3：

```
do
{
    do
    {
    ...
    } whiel();
}whiel();
```

形式 4：

```
do
{
    for(;;)
    {
    ...
    }
}
```

形式 5：

```
for(;;)
{
    while()
    {
    ...
    }
}
```

形式 6：

```
while()
{
    do
    {
    ...
    }while();
}
```

Form 2：

```
for(;;)
{
    for(;;)
    {
    ...
    }
}
```

Form 3：

```
do
{
    do
    {
    ...
    } whiel();
}whiel();
```

Form 4：

```
do
{
    for(;;)
    {
    ...
    }
}
```

Form 5：

```
for(;;)
{
    while()
    {
    ...
    }
}
```

Form 6：

```
while()
{
    do
    {
    ...
    }while();
}
```

【例3.3】显示如下的下三角九九乘法表。

```
1
2  4
3  6  9
4  8  12 16
5  10 15 20 15
6  12 18 24 30 36
7  14 21 28 35 42 49
8  16 24 32 40 48 56 64
9  18 27 36 45 54 63 72 81
```

【解题思路】该乘法表要列出 $1×2,2×1,2×2,3×1,3×2,3×3,\cdots,9×9$ 的值,乘数的范围是1至9,针对每一个乘数,被乘数的范围是1至它本身,因此,可以使用两重循环解决问题,按乘数组织外层循环,i 表示从1至9;按被乘数组织内层循环,j 表示从1至i,从而确定每一行输出的内容。

参考程序如下:

【Example 3.3】The multiplication table in triangular arrangement has been presented as following.

```
1
2  4
3  6  9
4  8  12 16
5  10 15 20 15
6  12 18 24 30 36
7  14 21 28 35 42 49
8  16 24 32 40 48 56 64
9  18 27 36 45 54 63 72 81
```

【Problem-Solving Ideas】The values of $1×2,2×1,2×2,3×1,3×2,3×3,\cdots,9×9$ should be listed in this multiplication table, the range of multiplier is from 1 to 9, aiming at each multiplier, the range of multiplicand is from 1 to itself, therefore, the double loop can be utilized for solving problems, according to multiplier organization to perform external loop, i means 1 to 9; According to multiplicand organization to perform internal loop, j means from 1 to i, so as to determine the output content of each line.

The referential program is as following:

```c
#include <stdio.h>
main()
{
    int i,j;
    for(i=1;i<=9;i++)
    {
        for(j=1;j<=i;j++)
        {
            printf("%-5d",i*j);
        }
        printf("\n");
    }
}
```

【例3.4】有一老大爷去集贸市场买鸡,他想用100元钱买100只鸡,而且

【Example 3.4】There is an old man goes to pedlars' market for buying chickens,

要求所买的鸡有公鸡、母鸡、小鸡。已知公鸡 2 元一只，母鸡 3 元一只，小鸡 0.5 元一只。问老大爷要买多少只公鸡、母鸡、小鸡恰好花去 100 元钱，并且买到 100 只鸡？请用双循环实现。

【解题思路】假设公鸡买 x 只，母鸡买 y 只，小鸡买 z 只，则

（1）y 可以是 1,2,3,…,33 的一个值。

（2）x 可以是 1,2,3,…,50 的一个值。

（3）根据 y 和 x，可以推导出 z=100-y-x。

（4）如果所花的钱刚好是 100，则输出 x,y,z。参考程序如下：

he wants to buy a hundred chickens with one hundred yuan, which including roosters, hens and chicks. It is known that one rooster is 2 yuan, one hen is 3 yuan, and one chick is 0.5 yuan. The question is that how many chickens should the old man buy for spending 100 yuan exactly, and the amount of chickens bought is one hundred? Please realize it by double loop.

【Problem-Solving Idea】Supposing to buy x roosters, y hens and z chicks, then

(1)y can be a value among 1, 2, 3, …, 33.

(2)x can be a value among 1, 2, 3, …, 50.

(3)In light of y and x, it can be deduced that z=100-y-x.

(4)If the money spent is exactly 100, then outputting x, y and z. The referential program is as following:

```c
#include <stdio.h>
main()
{
    int x,y,z;
    for(x=1;x<=50;x++)
        for(y=1;y<=33;y++)
        {
            z=100-x-y;
            if(2*x+3*y+0.5*z==100) printf("公鸡数为%d,母鸡数为%d,小鸡数为%d\n",x,y,z);
        }
}
```

3.3 任务实现

3.3.1 求一个小组学生成绩的总分及平均分

现要输入第一小组学生（人数为 8 人）的成绩，计算这一小组的总分与平均分，

3.3 Task Implementation

3.3.1 Calculating the Total and Average Scores of Students in One Group

Currently, inputting the scores of students in the first group (the number of people is eight), calculating the total and

并按要求输出。

小组有8个同学，显然需要使用循环的方式来进行，可以使用3种循环结构（while 循环、do...while 循环和 for 循环）中的任意一种实现。

实现参考代码如下：

average scores of students in this group, and outputting conforming to the requirements.

There are eight students in one group, it is clear that the loop method should be utilized for performing, and one of the three kinds of loop structures can be utilized (while loop, do...while loop and for loop).

Realizing referential code:

```c
#include <stdio.h>
main()
{
    int i;
    float x,y,z,sum,avg;
    i=1;
    while(i<=8)
    {
        printf("请输入第%d个同学三门课的成绩",i);
        scanf("%f%f%f",&x,&y,&z);
        sum=x+y+z;
        avg=sum/3;
        printf("第%d个同学的总分为%.2f,平均分%.2f\n",i,sum,avg);
        i=i+1;
    }
}
```

3.3.2 求一个小组学生成绩的总分及平均分

现要输入全班4个小组（假设每个小组10人）的学生成绩，计算每个小组的总分与平均分，并按要求输出。

在任务3.3.1中，所解决的问题是一个小组学生成绩的总分及平均分。现在则有4个小组，显然我们可以使用循环的嵌套来实现4组学生成绩的输入。

3.3.2 Calculating the Total and Average Scores of Students in One Group

Currently, inputting the student scores of students in four groups in the whole class (supposing that each group has ten people), calculating the total and average scores of students in each group, and outputting conforming to the requirements.

In task 3.3.1, the solved problem is the total and average scores of students in one group. Now there are four groups, it is clear that the loop embedding can be utilized for inputting the student scores in four groups.

Third Task: The Grouping and Summarizing of Student Scores

```c
1  #include "stdio.h"
2  main()
3  {
4      int score,i,sum;
5      float avg;
6      int j=1;
7      for(;j<=4;j++)
8      {
9          sum=0;
10         printf(" 请输入第%d小组学生成绩:",j);
11         for(i=1;i<=10;i++)
12         {
13             scanf("%d",&score);
14             sum=sum+score;
15         }
16         avg=sum/10.0;
17         printf("本小组10个学生的总分为：%d\n",sum);
18         printf("本小组10个学生的平均分为：%.2f\n",avg);
19     }
20 }
```

程序运行结果如图 3.6 所示。

The operation result is presented in Figure 3.6.

图 3.6　程序运行结果
Figure 3.6　Program Operation Result

【练习与提高】

1.编程:求 1+2！+3！+…+10!的和，要求用双循环的方法解决。

2.编程：用循环输出下面的图形。
*
***
*****
****

【Practice and Improvement】

1. Programming: Calculating the summation of 1+2!+3!+…+10!, it is required to utilize double loop method for solving.

2. Programming: Applying loop to output the following figure.
*
***
*****
******

3. 小明所在的部门一共分 4 个小组，每个小组人数不等，一般为 8～10 人，输入每个小组的人数及每个员工的工资，求每个小组员工的平均工资。

4. 求 1 至 100 的偶数的和。

5. 求 1–3+5–7+…–99 之和。

6. 输入两个整数，输出它们的最大公倍数。

7. 程序功能：对 x=1,2,3,…,10，求 $f(x)=x \cdot x-5 \cdot x+\sin x$ 最小值。

8. 程序功能：对 s=1+1/2+1/3+…直到加到最后一项的值小于 10^{-6} 为止。

3. There is a total of four groups in the department of Xiaoming, the number of people in each group is different, it is generally 8 to 10 people in one group, inputting the number of people for each group as well as the salary of each personnel, it is required to calculate the average salary of personnel in each group.

4. Calculating the summation of even numbers from 1 to 100.

5. Calculating the summation of 1–3+5–7+…–99.

6. Outputting two integers, outputting their greatest common multiple.

7. Program Function: for x=1,2,3,…,10, calculating the minimum value of $f(x)=x \cdot x-5 \cdot x+\sin x$.

8. Program Function: for s=1+1/2+1/3+…, until the summation of last item is less than 10^{-6}.

任务 4
学生成绩排序

Fourth Task: Sorting Student Scores

【知识目标】
1. 熟知一维数组的定义、存储及引用。
2. 了解二维数组的定义、存储及引用。

【能力目标】
1. 能够熟练应用数组解决实际问题。
2. 学会使用数组编写实用小程序。

【重点、难点】
1. 一维数组的理解及应用。
2. 二维数组的理解、引用及应用。

【课程思政】
1. 通过数组的定义，即具有相同的数据类型的数的集合，告诫学生物以类聚、人以群分，交友一定要慎重，必须有选择性地交友和学习，减少无效的社交。
2. 通过二维数组的定义，延伸对人生的思考，扩宽学生的思维。

【推荐教学方法】
通过教学做一体化教学，结合生活中常见的事例，使学生掌握知识点，学会编制程序流程图并进行程序的编写。

【推荐学习方法】
通过完成任务，在做中学、学中做，掌握实际技能与相关知识点。

【Knowledge Objective】
1. Familiar with the definition, storage and citation of one-dimensional array.
2. Having a good knowledge of the definition, storage and citation of two-dimensional array.

【Competency Objective】
1. Familiar with the application of arrays for solving real problems.
2. Learning to utilize arrays for compiling practical mini programs.

【Focal and Difficult Points】
1. The comprehending and application of one-dimensional array.
2. The comprehending, citation and application of two-dimensional array.

【Curriculum Ideological and Political Education】
1. By means of the definition of arrays, namely, the set of numbers with the same data type, educating the students that things of one kind come together, and people of one mind fall into the same group, they must be cautious in making friends, there must make friends and learn selectively, reducing the useless social relations.
2. By means of definition of two-dimensional arrays, expanding to the reflections on life, broadening the thinking of students.

【Recommended Teaching Method】
By means of integrated teaching, combining the common cases in daily life, making the students grasp knowledge points, learning to compile program flow diagram and edit program.

【Recommended Learning Method】
By means of accomplishing tasks, learning by doing, doing by learning, mastering the actual technologies and relevant knowledge points.

4.1 任务描述

班级的同学参加完期终考试（考了三门课），现在需要按成绩的高低输出成绩单。程序要求：成绩分数为 0～100 分，任意输入，为了方便，假设只有 5 位同学，程序运行结果如图 4.1 所示。

4.1 Task Description

The students in class have accomplished the final examination (three lessons), now it is necessary to output the scripts by descending order. Program requirements: the scores are from 0 to 100 points, inputting randomly for convenience, supposing that there are only five students, the program operation was presented in Figure 4.1.

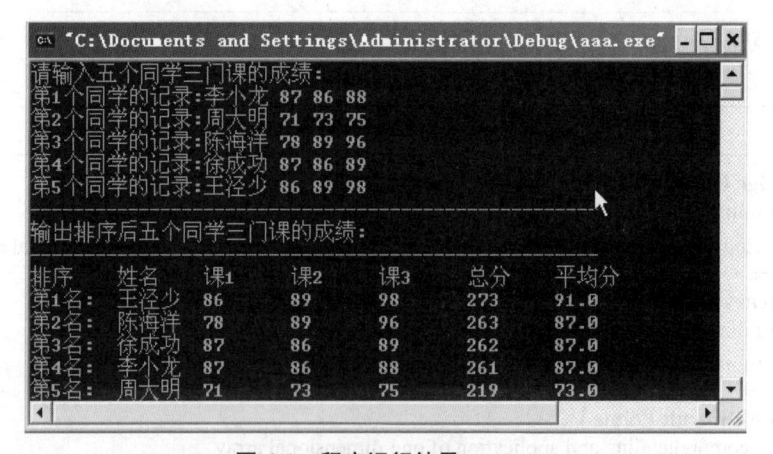

图 4.1　程序运行结果

Figure 4.1　Program Operation Result

4.2 相关知识

4.2.1 一维数组

1. 一维数组的定义

一维数组的定义方式为：

类型说明符 数组名 [常量表达式]

例如：

int a[20];

它表示数组名为 a，该数组的长为 20 个元素，每个元素均为 int 类型。

注意：

（1）数组名等同变量名，命名规则与变量名一样。

（2）数组中的常量表达式表示数组的长度，数组一经定义，长度就固定不变，因此用方括号括起来的可以是常量表达式、符号常量，不可能是变量。

（3）数组的下标从 0 开始。如 int a[20]; 表示定义了 20 个数组元素，分别为 a[0]，a[1]，a[2]，…，a[19]。若要引用第 i 个元素，则可以表示成 a[i]。

（4）数组名不能与其他变量名相同。

2. 一维数组的引用

数组元素的表示方式：数组名 [下标]，下标可以是常量、表达式、变量，如：a[5]、a[5-3]、a[i]。

【例 4.1】输入 10 个学生的成绩，并

4.2 Relevant Knowledge

4.2.1 One-Dimensional Array

1. The Definition of One-Dimensional Array

The definition method of one-dimensional array:

Type specifier Array name [Constant quantity expression]

For example

int a[20];

It signifies that the array name is a, the length of this array is twenty elements, and each element is of int type.

Attention:

(1) The array name is equivalent to variable name, the nomination rule is equivalent to variable name.

(2) The variable expression in array signifies the length of array, since the array is defined, the length will be fixed, therefore, those within square brackets can be variable expressions and character constant quantities, they cannot be variables.

(3) The subscript of array starts with 0. Such as, int a[20]; Signifying that 20 array elements have been defined, they are a[0], a[1], a[2], …, a[19]. If the ith element needs to be cited, then it can be presented as a[i].

(4) The array name cannot be the same as other variable names.

2.The Citation of One-Dimensional Array

The expression of array elements: Array name [subscript], subscript can be constant quantity, expressions and variables, such as: a[5], a[5-3] and a[i].

【Example 4.1】Inputting the scores of

将其输出。

【解题思路】假定学生的成绩都是整数，则需要定义一个长度为10的整形数组，使用循环的方式输入数据，并存在数组中，然后再通过循环的方式按序输出数组元素。

参考代码如下：

ten students, and then outputting them.

【Problem-Solving Ideas】Supposing that the students scores are all integers, then it is necessary to define the integer arrays with length of ten, applying loop method to input data and saving it in arrays, then outputting array element by means of loop method.

The referential code is as following:

```c
#include "stdio.h"
main()
{
    int i,a[10];
    printf("输入数组元素：");
    for(i=0;i<10;i++)
        scanf("%d",&a[i]);
    printf("输出数组元素：");
    for(i=0;i<10;i++)
        printf("%5d",a[i]);
}
```

3. 一维数组的初始化

对数组元素的初始化方法主要有以下几种方式：

（1）定义数组时，直接对数组元素赋以初值。例如：

int x[5]={0,1,2,3,4};

（2）只对一部分元素赋初值。例如：

int x[5]={0,1};

系统将自动给指定的数组元素赋值：x[0]=0,x[1]=1,其他元素值均为0。

（3）如果一个数组的全部元素值都为1，可以写成：

int x[5]={1,1,1,1,1}; 或者 int x[5]={1};

（4）对全部元素赋初值时，可以不指

3. The Initialization of One-Dimensional Array

There are several methods for initializing the array elements:

(1)When defining array, assigning the initialized value to array element directly. For example

int x[5]={0,1,2,3,4};

(2)Only assigning the initialized value to some elements. For example

int x[5]={0,1};

The system would assign the value to designated array elements automatically:x[0]=0,x[1]=1, the value of other elements can also be 0.

(3)If the values for all of the elements in a array are 1, then it can written as:

int x[5]={1,1,1,1,1};or int x[5]={1};

(4)When assigning the initialized value

Fourth Task: Sorting Student Scores

定长度。例如：

　　int x[5]={0,1,2,3,4}; 等价于 int[] = {0,1,2,3,4};

【例4.2】求本班同学的最高分，并将它与第一位数互换。

【解题思路】首先输入10个数给math[0]至math[9]（为了程序运行方便，假设只有10个同学），第二步是依次将math[0]与各个数进行比较，如果math[0]<math[i],则两数进行交换。经过上述过程后的math[0]就是最高分了。

参考程序如下：

to all of the elements, the length don't need to be designated. For example

　　int x[5]={0,1,2,3,4}; equivalent to int[] = {0,1,2,3,4};

【Example 4.2】Calculating the highest score of students in the whole class, and changing them with the first digit number.

【Problem-Solving Ideas】First inputting 10 numbers to math[0] to math[9] (for the convenient operation of program, supposing that there are only ten students), the second step is to compare math[0] with each number in sequence, if math[0]<math[i], then changing these two numbers. The math[0] experiencing the above procedures is the highest score.

The referential program is as following:

```c
#include "stdio.h"
main()
{
    int i,math[10],t;
    printf(" 请输入本班同学的成绩: ");
    for(i=0;i<10;i++)
        scanf("%d",&math[i]);
    for(i=1;i<10;i++)
        if(math[0]<math[i])
            {t=math[0]; math[0]=math[i]; math[i]=t;}
    printf("本班同学中的最高分: ");
    printf("%d\n", math[0]);
}
```

【例4.2】程序执行结果如下：

The operation result of【Example 4.2】program is as following:

```
请输入本班同学的成绩: 76 77 67 88 91 72 83 82 83 81
本班同学的最高分: 91
```

图4.2　【例4.2】程序运行结果
Figure 4.2　【Example 4.2】Program Operation Result

【例4.3】多个学生一门成绩的排序（比较法）。

【Example 4.3】The ranking for scores of one subject of several students (comparison method).

67

【解题思路】在【例4.2】中已求出了多个学生成绩的最高分，显然，按【例4.2】的逻辑，在剩下的数中再找出最高分，则将是多个学生的次高分，交换数据的算法为

```
for(i=2;i<10;i++){
    if(math[1]<math[i]){
        t=math[1];math[1]=math[i];math[i]=t;
    }
}
```

假设原先有10个数，则重复9次就可以达到排序的目的，也就是再嵌套一个循环，所以多个学生的成绩排序算法可以写成：

【Problem-Solving Ideas】The highest scores of several students have been calculated in【Example 4.2】, clearly, according to the 【Example 4.2】logic, finding the highest scores in the rest numbers again, then it will be the secondary highest scores for several students, the algorithm for changing data is:

```
for(i=2;i<10;i++){
    if(math[1]<math[i]){
        t=math[1];math[1]=math[i];math[i]=t;
    }
}
```

Supposing that there are 10 numbers at the beginning, then the ranking purpose can be achieved by repeating 9 times, that is to embed another loop again, therefore, the ranking algorithm of scores for several students can be written as:

```
1  #include "stdio.h"
2  main()
3  {
4      int i,math[10],t,j;
5      printf(" 请输入多个同学的成绩：");
6      for(i=0;i<10;i++)
7          scanf("%d",&math[i]);
8      for(j=0;j<9;j++)         //循环九次，就可以分离出前九个数
9          for(i=j+1;i<10;i++)  //分离第j个数，则一定与第j+1至最后一个数比较
10             if(math[j]<math[i])
11                {t=math[j]; math[j]=math[i]; math[i]=t;}
12     printf("多个同学的成绩排序为：");
13     for(i=0;i<10;i++)
14         printf("%3d", math[i]);
15     printf("\n");
16  }
```

以上代码的此种方法称为比较法，上述程序执行结果如下所示：

The method for above codes is comparison method, the implementation result for the above procedures is as following:

图4.3 【例4.3】程序运行结果
Figure 4.3 【Example 4.3】Program Operation Result

Fourth Task: Sorting Student Scores

【例 4.4】多个学生一门成绩的排序（用冒泡法排序）。

【解题思路】冒泡法，顾名思义，此算法像水中冒泡一样，水泡越大，则它浮出水面的速度就越快。思路是将两个数比较，将小的数调到后面，以简单的 18,9,1,2,6 按从大到小的顺序排序，方法为：

先将 18,9 比较，因为 18>9，所以 18 与 9 不交换，即还是 18,9,1,2,6；将 9 与 1 比较，因为 9>1，所以 9 和 1 不交换。即还是 18,9,1,2,6；将 1 与 2 比较，因为 1<2，所以 1 与 2 交换。即 18,9,2,1,6；将 1 与 6 比较，因为 1<6，所以 1 与 6 交换。即 18,9,2,6,1；通过以上 4 步，将最小的 1 沉在底部，而完成上述的程序可以表示成

```
for(i=0;i<4;i++)
  if(math[i]<math[i+1]) {t=math[i]; math[i]=math[i+1]; math[i+1]=t;}
```

接下来所要做的就是在剩下的数 18,9,2,6 中找出次小数，显然只要比较 3 次即可，即

```
for(i=0;i<3;i++)
  if(math[i]<math[i+1]) {t=math[i]; math[i]=math[i+1]; math[i+1]=t;}
```

依此类推，即可通过循环的嵌套实现排序，即

【Example 4.4】The ranking of scores for one subject of several students (ranking by bubbling method).

【Problem-Solving Ideas】The bubbling method, just as its name implies, this algorithm is like the bubbles in water, the bigger the bubble is, the faster it will emerge from the water. The idea is to compare these two numbers, adjusting the minimum number to the latter, utilizing the simple 18,9,1,2,6 to rank by descending order, the method is:

First comparing 18 and 9, because 18>9, then 18 and 9 can not be changed, that is 18,9,1,2,6 again; Comparing 9 and 1, because 9>1, then 9 and 1 can not be changed. That is 18,9,1,2,6 again; Comparing 1 and 2, because 1<2, then 1 and 2 can be changed. That is 18,9,2,1,6; Comparing 1 and 6, because 1<6, so 1 and 6 can be changed. That is 18,9,2,6,1; By means of the above four steps, sinking the minimum 1 to the bottom, while the programs that accomplishing the above procedures can be presented as

```
for(i=0;i<4;i++)
  if(math[i]<math[i+1]) {t=math[i]; math[i]=math[i+1]; math[i+1]=t;}
```

The next is to find the secondary minimum number among the rest numbers of 18,9,2,6, clearly, it only needs to compare them for three times, that is

```
for(i=0;i<3;i++)
  if(math[i]<math[i+1]) {t=math[i]; math[i]=math[i+1]; math[i+1]=t;}
```

And so forth, the ranking can be realized by loop embedding, namely:

for(j=0;j<4;j++)
for(i=0;i<4-j;i++)
if(math[i]<math[i+1]){t=math[i]; math[i]=math[i+1];
math[i+1]=t;}

综上所述，参考代码如下：

for(j=0;j<4;j++)
for(i=0;i<4–j;i++)
if(math[i]<math[i+1]){t=math[i]; math[i]=math[i+1];
math[i+1]=t;}

Above all, the referential code is as following:

```
#include "stdio.h"
main()
{
    int i,math[10],t,j;
    printf(" 请输入多个同学的成绩：");
    for(i=0;i<10;i++)
        scanf("%d",&math[i]);
    for(j=0;j<9;j++)
        for(i=0;i<9-j;i++)
            if(math[i]<math[i+1])
                {t=math[i]; math[i]=math[i+1]; math[i+1]=t;}
    printf("多个同学的成绩排序为：");
    for(i=0;i<10;i++)
        printf("%3d", math[i]);
    printf("\n");
}
```

【例 4.4】程序运行结果如图 4.4 所示。

The program【Example 4.4】operation result can be presented as Figure 4.4:

图 4.4 【例 4.4】程序运行结果

Figure 4.4 【Example 4.4】Program Operation Result

4.2.2 一维字符数组

1. 一维字符数组的定义

定义方法同数值型数组，例如：

char c[10];

以上是定义了一个字符数组 c，它有 10 个元素。

4.2.2 One-Dimensional Character Array

1. The Definition of One-Dimensional Character Array

The definition method is the same as that of numeric type array, for example:

char c[10];

The above is regarding the definition of one character array c, it has ten elements.

2. 一维字符数组的引用

（1）用 %c 格式符逐个输入/输出。

例如：

```
char c[6];
for(i=0;i<6;i++){
    scanf( "%c" ,&c[i]);
    printf( "%c" ,c[i]);
}
```

（2）用 %s 格式符进行字符串的输入/输出。

```
char c[6];
scanf( '%s' ,c);
printf( '%s' ,c);
```

注意：

（1）输出时，遇 '\0' 结束，且输出字符中不包含 '\0'。

（2）"%s" 格式符输入时，遇空格或回车结束，但获得的字符中不包含回车及空格本身，而是在字符末尾添加 '\0'。如

```
char c[10];
scanf( "%s" ,c);
```

如输入数据 "How are you"，结果仅 "How" 被输入到数组 c 中。

（3）一个 scanf( ) 函数可以输入多个字符串，输入时以空格键作为字符串间的分隔。例如：

```
char s1[5],s2[5],s3[5];
scanf( "%s%s%s" ,s1,s2,s3);
```

如输入数据 "How are you"，s1,s2,s3 获得的数据如下：

2. The Citation of One-Dimensional Character Array

(1) Applying %c format character to input/output one by one. For example

```
char c[6];
for(i=0;i<6;i++){
    scanf( "%c" ,&c[i]);
    printf( "%c" ,c[i]);
}
```

(2)Applying %s format character to input/output character strings.

```
char c[6];
scanf( "%s" ,c);
printf( "%s" ,c);
```

Attention:

(1) While outputting, it will end when encountering '\0', and the outputting character doesn't contain '\0'.

(2) While inputting "%s" format character, it will end encountering blank or carriage return, but the obtained character doesn't contain carriage return or the blank itself, but to add '\0' at the end of characters. For example:

```
char c[10];
scanf( "%s" ,c);
```

For example, when inputting data "How are you", as a result, only the "How" will be input in arrays.

(3) Several character strings can be input by one scanf function, while inputting, utilizing the blank button as the separator among character strings. For example:

```
char s1[5],s2[5],s3[5];
scanf( "%s%s%s" ,s1,s2,s3);
```

For example, when inputting data "How are you", the obtained data for s1,s2,s3 is as following:

s1:	H	o	w	\0	\0
s2:	a	r	e	\0	\0
s3:	y	o	u	\0	\0

（4）"%s"格式符输出时，若数组中包含一个以上'\0'，遇第一个'\0'时结束。

3．一维字符数组的初始化

对字符数组元素的初始化方法主要有以下几种方式：

（1）定义字符数组时，直接对数组元素赋以初值。例如：

char x[5]={ 'c' , 'h' , 'I' , 'n' , 'a' };

（2）可以省略数组长度。例如：

char x[]={ 'c' , 'h' , 'I' , 'n' , 'a' };

系统将根据初始值个数确定数组的长度，长度为5。

（3）字符数组可以用字符串来初始化，定义如下：

char x[6]= "china" ;

（4）可以省略数组长度。例如：

char x[]={ 'c' , 'h' , 'I' , 'n' , 'a' };

4．常用的字符串处理函数

（1）输入字符串函数：gets()。功能：从键盘输入一个字符串，允许输入空格，示例：

(4) While outputting "%s" format character, if the array contains one or more '\0', then it will end encountering the first '\0'.

3. The Initialization of One-Dimensional Character Array

There are primarily the following several methods for initializing the character array elements:

(1) When defining the character arrays, directly assigning the initialized value to array elements. For example:

char x[5]={ 'c' , 'h' , 'I' , 'n' , 'a' };

(2) The array length can be omitted. For example:

char x[]={ 'c' , 'h' , 'I' , 'n' , 'a' };

The system would decide the length of arrays according to the number of initialized value, the length is 5.

(3) The character array can be initialized by character strings, the definition is as following:

char x[6]= "china" ;

(4) The array length can be omitted. For example:

char x[]={ 'c' , 'h' , 'I' , 'n' , 'a' };

4. The Commonly Utilized Character Strings for Dealing with Function

(1) Inputting the character string function: gets(). Function: Inputting one character string from keyboard, allowing to input blank, for example

Fourth Task: Sorting Student Scores

```
char s[20];
gets(s);
```

(2) Outputting character string function: puts (). Function: Outputting the character strings saved in character arrays to standard outputting device, and replacing '\0' with '\n'. For example

```
char s[6] = "china";
puts(s);
```

【Example 4.5】By means of initialized method, storing the math scores of ten students in the learning group of Xiaogang in arrays, then inputting one examination score from keyboard, searching for this number in the arrays, if it can be found, then outputting the result regarding the order of student who obtaining this score in group.

【Problem-Solving Ideas】Defining and initializing the array a[10], then defining the variable k and storing the score input from keyboard, comparing k with the element in arrays accordingly, if they are the same, then adding its subscript value to 1 to obtain the order of student in group, exiting the loop.

The referential program is as following:

```c
#include <stdio.h>
void main()
{
    int i;
    int a[10]={78,87,91,65,75,82,74,85,78,70};
    int k;
    printf("请输入要查找的考分");
    scanf("%d",&k);
    for (i=0;i<10;i++)
        if (k==a[i])
        {
            printf("学生在小组中的排列位置是:%d\n",i+1);
            break;
        }
    if(i==10) printf("对不起，没有分数一致的人\n");
}
```

【例 4.6】三个同学姓名的输入输出。

【解题思路】定义三个字符数组，分别用来存三个姓名，参考代码如下：

【Example 4.6】The inputting and outputting of names for three students

【Problem-Solving Ideas】Defining three character arrays, storing the three names respectively, the referential code is as following:

```
#include "stdio.h"
main()
{
    char name1[10],name2[10],name3[10];
    printf("请输入姓名:\n");
    scanf("%s%s%s",name1,name2,name3);
    printf("输出的姓名为:\n");
    printf("%s,%s,%s\n",name1,name2,name3);
}
```

【例 4.7】将【例 4.6】改为 gets() 输入。

【Example 4.7】Changing【Example 4.6】to gets() to input.

```
#include "stdio.h"
main()
{
    char name1[10],name2[10],name3[10];
    printf("请输入姓名:\n");
    gets(name1);
    gets(name2);
    gets(name3);
    printf("输出的姓名为:\n");
    printf("%s,%s,%s\n",name1,name2,name3);
}
```

4.2.3 二维数组

1. 二维数组的定义

二维数组定义的一般形式为

类型说明符 数组名 [常量表达式][常量表达式]

4.2.3 Two-Dimensional Array

1. The Definition of Two-Dimensional Array

The general format for two-dimensional array definition is:

Type specifier Array name [constant quantity expression][constant quantity expression]

例如：int a[3][4]，它表示数组名为a，该数组的长为12个元素，定义了一个3行4列的整型数组。

由上面定义可以看出，二维数组的实质是一种特殊的一维数组，即一维数组中的每个元素都是一个一维数组的数组。例如上述的例子 int a[3][4]，可以看成是一个具有3个元素的特殊一维数组，三个元素分别为 a[0]、a[1] 和 a[2]，其中 a[0] 里面存放的数据是一个具有4个元素的一维数组，如 a[1] 这个元素包含4个元素，分别为 a[1][0]、a[1][1]、a[1][2] 和 a[1][3]。

2. 二维数组的引用

二维数组元素的表示形式为

数组名 [下标][下标]

例如，int a[3][4]，表示行下标值从0开始，最大为2，列下标从0开始，最大为3，各个元素可表示为以下的二维表：

a[0][0] a[0][1] a[0][2] a[0][3]
a[1][0] a[1][1] a[1][2] a[1][3]
a[2][0] a[1][1] a[2][2] a[2][3]

3. 二数组的初始化

二维数组的初始化方法如下：

（1）分别给各维度赋初值。

For example int a[3][4], it means that the array name is a, the length of this array is twelve elements, defining the integer type array with three lines and four columns.

It can be observed from the above definition that the nature of two-dimensional array is a special one-dimensional array, namely, each element in one-dimensional array is the array for one-dimensional array. For example, as for the above example int a[3][4], it can be taken as the special one-dimensional array composed of three elements, the three elements are a[0], a[1] and a[2] respectively, among them, the data stored in a[0] is a one-dimensional array composed of four elements, for example, the element of a[1] contains four elements, they are a[1][0], a[1][1], a[1][2] and a[1][3] respectively.

2. The Citation of Two-Dimensional Array

The expression form of two-dimensional array elements is:

Array name [subscript][subscript]

For example, int a[3][4] means that the subscript value below line starts from 0, the greatest value is 2, listing the subscript value to start from 0, the greatest value is 3, each element can be presented as the following two-dimensional table:

a[0][0] a[0][1] a[0][2] a[0][3]
a[1][0] a[1][1] a[1][2] a[1][3]
a[2][0] a[1][1] a[2][2] a[2][3]

3. The Initialization of Two-Dimensional Array

The initialization method of two-dimensional array is as following:

(1) Assigning the initialized value to

例如：int a[3][4] = {{1,2,3,4},{2,3,4,5},{3,4,5,6},{4,5,6,7}};

（2）将所有数据写在一个花括号内，按数值排列的顺序对各元素赋初值。

例如：int a[3][4] = {1,2,3,4,5,6,7,8,9,10,11,12};

（3）允许只对部分元素赋初值。

例如：int a[3][4] = {{1,2,3},{5},{1,2,5}};
则 a 得到的数组为

$$\begin{pmatrix} 1 & 2 & 3 & 0 \\ 5 & 0 & 0 & 0 \\ 1 & 2 & 5 & 0 \end{pmatrix}$$

（4）如果对全部数组元素赋值，则第一维的长度可以不指定，但必须指定第二维的长度，全部数据写在一个大括号内。

例如：int a[][4] = {1,2,3,4,5,6,7,8,9,10,11,12};

$$\begin{pmatrix} 1 & 2 & 3 & 4 \\ 5 & 6 & 7 & 8 \\ 9 & 10 & 11 & 12 \end{pmatrix}$$

【例 4.8】输入五个同学三门课的成绩并输出。

【解题思路】根据二维数组的定义和

each dimension.

For example int a[3][4] = {{1,2,3,4},{2,3,4,5},{3,4,5,6},{4,5,6,7}};

(2) Writing all of the data within one brace, assigning the initialized value to each element in accordance with the ranking order of arrays.

For example int a[3][4] = {1,2,3,4,5,6,7,8,9,10,11,12};

(3) Allowing to assign the initialized value to some elements.

For example int a[3][4] = {{1,2,3},{5},{1,2,5}};Then the array obtained by a is:

$$\begin{pmatrix} 1 & 2 & 3 & 0 \\ 5 & 0 & 0 & 0 \\ 1 & 2 & 5 & 0 \end{pmatrix}$$

(4) If assigning value to all of the array elements, then the one-dimensional length doesn't need to be designated, but the two-dimensional length must be designated, all of the data should be written within one big brace.

For example int a[][4] = {1,2,3,4,5,6,7,8,9,10,11,12};

$$\begin{pmatrix} 1 & 2 & 3 & 4 \\ 5 & 6 & 7 & 8 \\ 9 & 10 & 11 & 12 \end{pmatrix}$$

【Example 4.8】Inputting the scores for three subjects of five students and inputting them.

【Problem-Solving Ideas】In

引用，本题的思路参考代码如下：

accordance with the definition and citation of two-dimensional arrays, the idea referential code of this problem is as following:

```c
#include "stdio.h"
#define N 5
main()
{
    int i,j;
    int score [N][3];
    printf("请输入五个同学三门课的成绩:\n");
    for (i=0;i<N;i++)
        for(j=0;j<3;j++)
            scanf("%d",&score[i][j]);
    printf("输出五个同学三门课的成绩:\n");
    for(i=0;i<N;i++)
    {
        printf("第%d位同学:",i+1);
        for(j=0;j<3;j++)
            printf("%5d",score[i][j]);
        printf("\n");
    }
}
```

4.2.4 二维字符数组

1. 二维字符数组的定义

char str[3][5]; 表示定义一个二维字符数组 str，共有 3 行 5 列共 15 个元素。

2. 二维字符数组的初始化

二维字符数组的初始化方法如下：

（1）分别给各元素赋值。

例如：

charstr[3][4]={{'a','b','c','d'},{'e','f','g','h'},{'1','3','5','7'}};

数组形式如下所示：

$$\begin{pmatrix} a & b & c & d \\ e & f & g & h \\ 1 & 3 & 5 & 7 \end{pmatrix}$$

4.2.4 Two-Dimensional Array

1. The Definition of Two-Dimensional Character Array

char str[3][5]; Signifying to define one two-dimensional character array str, there is a total of 15 elements with three lines and five columns.

2. The Initialization of Two-Dimensional Character Array

The initialization method for two-dimensional character array is as following:

(1) Assigning value to each element respectively.

For example

charstr[3][4]={{'a','b','c','d'},{'e','f','g','h'},{'1','3','5','7'}};

The array form is as following:

$$\begin{pmatrix} a & b & c & d \\ e & f & g & h \\ 1 & 3 & 5 & 7 \end{pmatrix}$$

（2）使用字符串赋值。

例如：char str[3][3] = { "abc123" };

3. 二维字符数组的引用

比如输入/输出二维数组中的第 i 行（假设 i=3），则有两种实现方法：

方法 1：

char name[10][20];

gets(name[3]);

puts(name[3]);

方法 2：

har name[10][20];

scanf("%s",name[3]);

printf("%s",name[3]);

【例 4.9】从键盘输入一串字符（以 Enter 键结束），统计字符数。

【解题思路】因为不知道会输入一串多长的字符，所以刚开始定义一个足够大的字符数组 char str[80]，然后从键盘输入的字符将会放在 str[0] 至 str[k] 中，而 str[k+1] 将存放回车，如下：

(2) Applying character string to assign value.

For example char str[3][3] = {"abc123"};

3. The Citation of Two-Dimensional Character Array

For example, input/output the ith line in two-dimensional array (supposing i=3), then there are two realization methods:

Method 1:

char name[10][20];

gets(name[3]);

puts(name[3]);

Method 2:

char name[10][20];

scanf("%s",name[3]);

printf("%s",name[3]);

【Example 4.9】Inputting one string character from keyboard (ending with Enter key), calculating the character number.

【Problem-Solving Ideas】Since the length of inputting string character can not be determined, so first defining a big enough character array char str[80], then putting the character input from keyboard in str[0] to str[0], while return key will stored in str[k+1]. As it was presented in the following table:

str[0]	...	str[k]	\0				

所以，从 str[0] 开始找，只要没找到 '\0'，则继续找下一个。

参考代码如下：

Therefore, starting from str[0] to search, if '\0' cannot be found, then continuing to search for the next one.

The referential code is as following:

```c
1  #include "stdio.h"
2  #define N 80
3  main()
4  {
5      char str[N],i,t;
6      printf("输入字符串\n");
7      gets(str);
8      i=0;
9      while(str[i]!='\0')
10         i++;
11     printf("字符数为%d",i);
12     printf("\n");
13 }
```

4.3 任务实现

4.3.1 多个学生一门课成绩的输入 / 输出

班级的 40 个同学都参加了数学考试，现要输入全班同学的成绩，并逆序输出。全班有 40 个同学，显然需要定义一个长度为 40 的数组来存储成绩，在输出时，按数组的索引下标从大到小进行输出。

参考代码如下：

4.3 Task Implementation

4.3.1 The Input/Output of Scores for One Subject of Several Students

The forty students in class have participated in math examination, currently, inputting the scores of students in whole class, and ranking their scores by ascending order. There is a total of forty students in the class, clearly, it is necessary to define a array with the length of 40 to store the scores, outputting them according to the index subscripts by descending order.

The referential code is as following:

```c
1  #include "stdio.h"
2  main()
3  {
4      int i,score[40];
5      printf("请输入本班同学的成绩：");
6      for(i=0;i<40;i++)
7          scanf("%d",&score[i]);
8      printf("按逆序输出本班同学的成绩：");
9      for(i=39;i>=0;i--)
10         printf("%3d",score[i]);
11 }
```

4.3.2 多个学生一门课成绩的排序

班级的 40 个同学都参加了数学考试，

4.3.2 The Ranking of Scores for One Subject of Several Students

The forty students in class have participated in math examination, currently,

现要输入全班同学的成绩，并按学生成绩高低进行排序。

输入 40 个同学的数学成绩，在任务 4.3.1 中已经学会，只要定义一个数组 int math[40]，然后使用一个循环输入即可；对于学生成绩的排序，可以看成是求最高分的过程。假设 math[0] 为最高分，然后将 math[0] 与 math[i](i=1,2,…,39) 比较，如果 math[0]<math[i]，则将 math[0] 与 math[i] 交换；然后，在剩下的分数中求次高分，这样一直循环下去，直到将倒数第二个数找到为止。所以先要解决的是求多个同学的最高分问题，然后在剩下的分数中找次高分，不断重复，直到剩下的最后一个数是最小数为止。

inputting the scores of students in whole class, and ranking their scores by descending order.

Inputting the math scores of forty students, which have been learned in task 4.3.1, only to define one array int math[40], then utilizing one loop to input; as for the ranking of student scores, it can be taken as the process for calculating the highest score. Supposing that math[0] is the highest score, then comparing math[0] with math[i] (i=1,2,…,39), if math[0]<math[i], then changing math[0] with math[i]; after that, calculating the secondary highest score in the rest scores, continue to loop in this way, until the penultimate number is found. Therefore, first of all, it is necessary to solve the problem of calculating the highest score of several students, then searching for the secondary score in rest scores, repeating constantly, until the last number is the minimum number.

参考代码 1 如下（比较法）：

The referential code 1 is as following (comparison method):

```c
#include "stdio.h"
main()
{
    int i,math[10],t,j;
    printf(" 请输入多个同学的成绩：");
    for(i=0;i<10;i++)
        scanf("%d",&math[i]);
    for(j=0;j<9;j++)              //循环九次，就可以分离出前九个数
        for(i=j+1;i<10;i++)        //分离第j个数，则一定与第j+1至最后一个数比较
            if(math[j]<math[i])
                {t=math[j]; math[j]=math[i]; math[i]=t;}
    printf("多个同学的成绩排序为：");
    for(i=0;i<10;i++)
        printf("%3d", math[i]);
    printf("\n");}
```

参考代码 2 如下（冒泡法）：

The referential code 2 is as following (bubbling method):

Fourth Task: Sorting Student Scores

```c
1  #include "stdio.h"
2  main()
3  {
4      int i,math[10],t,j;
5      printf(" 请输入多个同学的成绩：");
6      for(i=0;i<10;i++)
7          scanf("%d",&math[i]);
8      for(j=0;j<9;j++)
9          for(i=0;i<9-j;i++)
10             if(math[i]<math[i+1])
11                 {t=math[i]; math[i]=math[i+1]; math[i+1]=t;}
12     printf("多个同学的成绩排序为：");
13     for(i=0;i<10;i++)
14         printf("%3d", math[i]);
15     printf("\n");
16 }
```

4.3.3 学生姓名的输入/输出

班级的 40 个同学都参加了数学考试，现在要求输入全班同学的姓名，并输出学生姓名。输入 40 个同学的成绩和多个学生一门课的成绩在 4.3.1 和 4.3.2 中已经学会，只需要掌握字符数组的输入/输出，定义一个二维字符数组，就能解决此问题。

参考代码如下：

4.3.3 The Input/Output of Student Names

The forty students in class have all participated in the math examination, currently, it is required to input the student names in class, and outputting the students names. Inputting the scores of forty students and the scores for one subject of several students have been learned in 4.3.1 and 4.3.2, it is only necessary to grasp the inputting/outputting of character arrays, this problem can be solved by defining only one two-dimensional character array.

The referential code is as following:

```c
1  #include "stdio.h"
2  #include "string.h"
3  #define N 5
4  main()
5  {
6      char name[N][12];
7      char tt[20];int i,j;
8      printf("请输入%d个候选同学的姓名:\n",N);
9      for(i=0;i<N;i++)
10         gets(name[i]);
11     printf("------------------\n");
12     printf("输出%d个候选同学的姓名:\n",N);
13     printf("------------------\n");
14     for(i=0;i<N;i++)
15         puts(name[i]);
16 }
```

4.3.4 多个学生多门课成绩的排序

班级的 40 个同学都参加了三门课程的考试，现在要求出按总成绩的高低排序的成绩单。成绩的格式如下：

排序	姓名	英语	数学	语文	总分	平均分
1	张三	80	68	78	××	××
2	李四	89	70	86	××	××

本问题要解决的多个学生姓名的输入/输出，这个任务在 4.3.3 中已经解决，同时也需要输入/输出 40 个同学的成绩，并进行相应的总分及平均分的计算，最后按总分的高低进行排序，所以可以将此问题分解为两个小任务，一个是用二维数组解决 40 个同学三门课的成绩输入/输出，另一个是计算相应的平均分及总分并进行排序。

参考代码如下（假设只有 5 名学生的情况）：

4.3.4 The Ranking of Multiple Subjects for Several Students

The forty students in class have all participated in the examinations of three subjects, currently, it is required to obtain the transcript rated by descending order. The format of scores is as following:

Serial Number	Name	English	Math	Chinese	Total Scores	Average Scores
1	Zhang San	80	68	78	××	××
2	Li Si	89	70	86	××	××

This problem aims to solve the inputting/outputting of several students names, this task has been solved in 4.3.3, in the meantime, it is necessary to input/output the scores of forty students, and carrying out the calculations of corresponding total scores and average scores of students, at last, ranking the total scores by descending order, therefore, this problem can be decomposed into two little tasks, one is to utilize two-dimensional array to solve the inputting/outputting of scores for three subjects of forty students, another is to calculate the corresponding average scores and rank.

The referential code is as following (supposing only to have the conditions of five students):

Fourth Task: Sorting Student Scores

```c
#include "stdio.h"
#include "string.h"
#define N 5
main()
{
    int i,j;
    int score [N][3],t;
    char name[N][10],nn[10];
    float sum[N]={0},avg[N];
    //每个同学的总分及平均分
    printf("请输入五个同学三门课的成绩:\n");
    /*输入记录*/
    for (i=0;i<N;i++)
    {
        printf("第%d个同学的记录:",i+1);
        scanf("%s",name[i]);
        for(j=0;j<3;j++)
            scanf("%d",&score[i][j]);
    }

    /*计算每个同学的总分与平均分*/
    for(i=0;i<N;i++)
    {
        for(j=0;j<3;j++)
            sum[i]=sum[i]+score[i][j];
        avg[i]=sum[i]/3.0;
    }
    /*排序成绩*/
    for(i=0;i<N-1;i++)
        for(j=0;j<N-1-i;j++)
            if(sum[j]<sum[j+1])
            {
                t=sum[j];sum[j]=sum[j+1];sum[j+1]=t;
                t=avg[j];avg[j]=avg[j+1];avg[j+1]=t;   //这个同学的所有数据都要交换
                t=score[j][0];score[j][0]=score[j+1][0];score[j+1][0]=t;
                t=score[j][1];score[j][1]=score[j+1][1];score[j+1][1]=t;
                t=score[j][2];score[j][2]=score[j+1][2];score[j+1][2]=t;
                strcpy(nn,name[j]);strcpy(name[j],name[j+1]);strcpy(name[j+1],nn);
            }

    printf("-------------------------------------------------\n");
    printf("输出排序后五个同学三门课的成绩:\n");
    printf("-------------------------------------------------\n");
    printf("排序\t姓名\t课1\t课2\t课3\t总分\t平均分\n");
    for (i=0;i<N;i++)
    {
        printf("第%d名:\t",i+1);
        printf("%s\t",name[i]);
        for(j=0;j<3;j++)
            printf("%d\t",score[i][j]);
        printf("%.0f\t%.1f\t",sum[i],avg[i]);
        printf("\n");
    }
    printf("-------------------------------------------------\n");
}
```

多个学生多门课成绩排序执行如图4.5所示。

The ranking of scores for multiple subjects of several students has been presented in the Figure 4.5.

图 4.5 程序运行结果
Figure 4.5 Program Operation Result

【练习与提高】

1. 输入张三所在部门 6 个同事的姓名，要求按逆序输出他们的花名册。

2. 读入一串数字，请统计其中数字 0～9 各数字出现的次数。

3. 输入 10 个数放在一维数组中，找出最小的数及其下标。

4. 从键盘上输入小明所在大组 20 个同学的成绩，输出他们的平均值及其中与平均值之差的绝对值为最小的那个同学的位置及分数。

5. 输出如下的杨辉三角，要求输出 5 行 5 列。

```
1
1  1
1  2  1
1  3  3  1
1  4  6  4  1
```

【Practice and Improvement】

1. Inputting the names for six colleagues in the department of Zhang San, it is required to output their membership roster by ascending order.

2. Reading one string of numbers, please calculate the emergency frequency of numbers from "0" to "9" among them.

3. Inputting ten numbers and putting them in one-dimensional array, searching out the minimum number and its subscript.

4. Inputting the scores of twenty students in the large group of Xiaoming from keyboard, outputting their average scores and the position and score of that student that having the minimum absolute value difference between score and average value.

5. Outputting the following Pascal's triangle, it is required to output five lines and columns.

```
1
1  1
1  2  1
1  3  3  1
1  4  6  4  1
```

Fourth Task: Sorting Student Scores

6. 在二维数组 A 中选出各行最大的元素组成一个一维数组 B。

7. 输入小明所在小组 6 个员工的姓名、资金、工资总和，并按工资总和从高到低排序输出。

8. 输入十个候选人同学的姓名，按 ASCII 码从大到小的顺序排序。

6. Selecting the greatest element in each line of two-dimensional array A to form a one-dimensional array B.

7. Inputting the summation of names, capitals and salaries for six personal in the group of Xiaoming, outputting and ranking in according to the salary summation by descending order.

8. Inputting the names for ten candidate students, ranking in accordance with ASCII codes by descending order.

任务 5
学生成绩分类汇总

Fifth Task: Categorizing and Summarizing Student Scores

【知识目标】 1. 熟练掌握函数的声明、定义和调用。 2. 能编写和调用无参数函数、有参函数。 3. 掌握函数的嵌套调用和递归调用。
【能力目标】 1. 能够熟练应用函数解决实际问题。 2. 学会使用局部变量和全局变量。
【重点、难点】 1. 有参函数形参和实参的值传递方式。 2. 函数的嵌套。
【课程思政】 1. 通过结构化的程序分析，培养学生工程项目分析能力和管理能力，同时加强学生的团队精神和团队合作能力。 2. 通过递归函数的定义，强调言传身教的重要性。
【推荐教学方法】 通过教学做一体化教学，结合生活中常见的事例，使学生掌握知识点，学会编制程序流程图并进行程序的编写。
【推荐学习方法】 通过完成任务，在做中学、学中做，掌握实际技能与相关知识点。

【Knowledge Objective】 1. Proficient in the statement, definition and invocation of functions. 2. Capable of compiling and invoking the parameterless function and parameter function. 3. Mastering the embedding invocation as well as the recursion invocation of function.
【Competency Objective】 1. Capable of applying functions proficiently to solve the actual problems. 2. Learning to utilize partial and overall variables.

Fifth Task: Categorizing and Summarizing Student Scores

【Focal and Difficult Points】
  1. The transmitting methods of values for formal parameter and actual parameter of parameter function.
  2. The embedding of function.

【Curriculum Ideological and Political Education】
  1. By means of structural program analysis, cultivating the project program analysis capabilities as well as the management capabilities of students, in the meantime, reinforcing the teamwork spirit and the teamwork cooperation capabilities of students.
  2. By means of defining the recursion function, stressing the importance of teaching by personal example as well as verbal instruction.

【Recommended Teaching Method】
  By means of integrated teaching, combining the common cases in daily life, making students master the knowledge points, learning to compile program flow diagram and edit programs.

【Recommended Learning Method】
  By means of accomplishing tasks, learning by doing, doing by learning, having a good command of actual technologies and related knowledge points.

5.1 任务描述

前面章节介绍了C语言的基本语句，已经可以进行程序设计了，但如果在编写较多子任务或解决较为复杂的实际问题时，程序代码会显得有些烦琐，同时编写类似代码会不可避免地使得程序变得冗长不简洁。如有这样一个任务，假设一个班有40位学生参加了期终考试，考了三门课，分别是C语言、高等数学、英语，40位学生分成5个组，每个组的学生人数可以不同，老师想统计以下信息：①统计小组一门课程的总分及平均分；②统计小组若干门课程的总分及平均分；③输出排序后小组三门课成绩单。该任务有多个子任务，程序代码必然冗长不简洁，严重缺乏可读性和可维护性，

5.1 Task Description

The basic statements of C language have been introduced in the former chapters, the program design can be performed, however, when compiling excessive subtasks or resolving relatively complicated actual problems, the program code can be a little complex, in the meantime, the programs can be lengthy and not simple due to the unavoidable compiling of similar codes. If there is such as task, supposing the forty students in a class have participated the final examinations for three subjects, including C language, superior mathematics and English, the forty students can be divided into five groups, the number of students in each group is not the same, the teacher wants to make statistics on the following information: ① making statistics on the total and average scores of one subject for groups; ② making

为了解决这一问题，可以将一个大的程序分成若干个程序模块，每一个模块用来实现一个特定的功能，这个程序模块就是函数。

statistics on the total and average scores of several subjects for groups; ③ Outputting the transcripts of three subjects for groups after ranking. This task contain several subtasks, the program code can be lengthy and not simple, be seriously short of readability and maintainability, for the purpose of solving this problem, a large program can be divided into several program modules, and each module can be utilized for realizing specific function, this program module is function.

5.2 相关知识

函数是 C 语言程序的基本组成单位，是用来完成独立的特定功能的程序模块，包含了若干条语句。一个 C 程序可由一个主函数 main() 和若干个其他函数构成，由主函数调用其他函数，其他函数也可以互相调用，同一个函数可以被一个或多个函数调用任意多次。

由于 C 语言采用函数模块的结构进行程序设计，因此 C 语言易于实现结构化程序设计，使得程序的层次结构清晰，便于程序的编写、阅读、调试。

5.2.1 函数的定义

函数包括标准库函数和自定义函数两种。库函数是由 C 语言系统提供定义的，在编程时用预编译命令 #include 将函数所

5.2 Relevant Knowledge

Function is the basic constituent unit of C language program, which is the program module utilized for accomplishing independent specific function, containing several statements. One C language program can be composed of one main function main() and several other functions, the main function would invoke other functions, other functions can also be invoked mutually, the same function can be invoked randomly by one or several functions.

Owing to the fact that C language applies the structure of function module to perform program design, therefore, C language can realize the structural program design easily, endowing the program with clear layers and structures, which is convenient for the programs to compile, read and invoke.

5.2.1 The Definition of Function

Functions are composed of standard library functions and customized functions. The definition of library function is provided by C language system, in programming,

在的头文件（*.h）进行引用即可，例如"stdio.h"头文件中包含了printf()函数、scanf()函数、getchar()函数、putchar()函数等。自定义函数是用户根据程序完成功能的需要而编写的程序。对于用于自定义函数，必须遵循先定义、后使用的规则，如果没有函数的定义，也就无法调用函数。

函数在调用时，会完成函数定义的特定功能，然后返回到调用它的地方，此时会有两种情况：①函数执行完毕后，会得到一个运算结果，并且把该结果送回到调用函数的地方，这种情况下的函数为有返回值函数；②函数只完成一系列的操作，并没有带回任何运算结果，或者不需要返回任何运算结果，这种情况下的函数为无返回值函数。

1. 有返回值函数的定义

有返回值函数定义的一般形式为

函数类型 函数名（形参列表） /* 函数首部 */
{
函数体语句； /* 函数体 */
}

函数类型是指函数返回值的数据类

applying the precompiling order #include to cite the heading file (*.h) of functions, for example, the heading file of "stdio.h" contains printf() function, scanf() function, getchar() function, putchar() function and some other functions. The customized function is the program compiled by users according to the requirements of program accomplishing function. For utilizing customized function, it is necessary to observe the rule of first definition and later application, if there is no definition for functions, then the functions cannot be invoked.

When invoking the function, the specific function defined by function can be accomplished, then returning to the invocation position, at this point, there will be two conditions: ① a operation result can be obtained after the function is performed, and the result would be sent to the position for invoking the functions, the function in such case is the function with returned values; ② the function only accomplishes a series of operations, no operation result has been obtained, or no operation result needs to be returned, in such case, the function constitutes the function without returned value.

1. The Definition of Function with Returned Value

The general form of definition for function with returned value is:

Function type Function name (formal parameter list)/*Function heading*/
{
Function body statement;   /*Function body*/
}

Function type refers to the data type of

型，一般与函数体内 return 语句表达式的数据类型一致。

函数名是用户自定义函数的名称，命名规则跟变量完全一样，为了增加程序的可读性，一般取有助于记忆并与其功能有关的名字作为函数名，在同一程序中，不能有同名的函数或变量。

形参是形式参数的简称，形参列表是由函数中所有形参名称和类型组成的列表，形参列表中的变量在调用之前没有实际的值，只是在形式上完成运算的一个替代符号，形参列表格式为

数据类型 1 形参 1，数据类型 2 形参 2，…，数据类型 n 形参 n

形参也可以一个都没有，称为无参函数。

函数体语句包括一对大括号和函数实现功能的所有语句，并用 return 语句返回运算结果。return 语句的一般格式为

return 表达式；

每个函数只能返回一个运算结果，如果函数体内有多个 return 语句，DEV C++ 编译器以执行第一条 return 语句中表

function returned value, which is generally the same as the data type of return statement expression in the function body.

The function name refers to the customized name defined by users, the nomination principle is completely the same as variable, for increasing the readability of programs, generally, applying the name which is conducive to memories and related to its function as the function name, in the same program, there cannot be the functions or variables with the same names.

FP is the short name for form parameter, formal parameter list is composed of all of the formal parameter names and types in functions, the variables in formal parameter list don't have actual values before invocation, which is only a alternative symbol utilized for accomplishing operation in formal sense, the format of formal parameter list is:

Data type 1 formal parameter 1, data type 2 formal parameter 2, ..., data type *n* formal parameter *n*.

It is also available if there is no formal parameter, in this case, it is called parameterless function.

Function body statement includes one pair of brace and all of the statements with function implementation capabilities, and applying return statement to return the operation result. The general format of return statement is:

return expression;

Each function can only return one operation result, if there are several return statements in function body, DEV C++ compiler should perform the value of

达式的值进行返回。

例如，定义一个函数，用于求两个数中的较大数，可将函数定义为

```
nt max(int x, int y)
{
    if(x>y)
    return x;
    else
    return y;
}
```

变量 a 和 b 都是 int 类型，因此 max() 函数返回值的类型和 return 语句所返回的值的类型应当要一致。

2. 无返回值函数的定义

无返回值函数定义的一般形式为

```
void 函数名（形参列表）    /* 函数首部 */
{
    语句;                /* 函数体 */
}
```

无返回值函数的函数类型为 void，表示只执行一个特定的功能，并不返回任何运算结果，函数体内一般也没有 return 语句。

例如，定义一个函数 star 输出一行 10 个 "*"，可将函数定义为

```
void star( )
{
    printf( "**********\n" );
}
```

expression in the first return statement and return.

For example, defining a function to calculate the greater value between two numbers, the function can be defined as

```
int max(int x, int y)
{
    if(x>y)
    return x;
    else
    return y;
}
```

The variable a and b are all of int type, therefore, the types of return values for max function should be consistent with that of return statement.

2. The Definition of Function without Return Values

The general format for definition of function without return value is:

```
void  Function name (formal parameter list)
                    /*Function heading*/
{
    Statement;   /*Function body*/
}
```

The type of function without return value is void, signifying to only perform one specific function, and don't return any operation results, there is usually no return statement can be found in function body.

For example, outputting one line of ten "*" to define one function star, the function can be defined as

```
void star( )
{
    printf( "**********\n" );
}
```

注意：

（1）当用户自定义函数的函数体语句只有一条语句时，大括号也必须保留。当用户自定义函数的函数体语句一句都没有时，此时的函数称为空函数，一般用来为后续预留功能扩展编写程序而占"位子"。

（2）无参函数是表示形参列表为空，但小括号是必须保留的。

（3）无返回值函数的 void 不能省略，否则函数返回值的类型默认为 int 型。

（4）所有的函数都是相互独立的、相互平行的，不存在上一级函数和下一级函数的问题，即不能在一个函数的函数体内定义另外一个函数。

5.2.2 函数的声明

函数声明的主要目的是在发生调用之前，向编译器说明函数的类型和参数，以保证程序编译时能判断对该函数的调用是否正确。

函数声明的一般格式为

    函数类型 函数名(形参列表);

函数声明是一条语句，后面要加上分号，即为函数定义的首部加上分号。

Attention:

(1) When there is only one statement for function body statement of customized function by user, the brace must be preserved. When there is no statement for function body statement of customized function by user, at this point, the function can be called as blank function, which is generally utilized for occupying "positions" for the latter reservation function to expand and compile programs.

(2) Parameterless function signifies that the formal parameter list is blank, but the parenthesis must be preserved.

(3) The void of function without return values can not be omitted, or else the type of function return value is taken as int type.

(4) All of the functions are mutually independent and parallel, there is no problems between the superior-level functions and inferior-level functions, namely, another function cannot be defined in one function body.

5.2.2 The Declaration of Function

The declaration of function aims to explain the function type and parameters to compiler before invocation, so as to assure that the invocation of this function can be judged in compiling programs.

The general format of function declaration is:

  Function type Function name (formal parameter list);

Function declaration is about a statement, which should be followed by semicolon, namely, adding semicolon to the heading of function definition.

Fifth Task: Categorizing and Summarizing Student Scores

C语言要求函数要先定义再调用，如果把自定义函数的定义放在主函数后面，那么必须在主函数前面对函数进行声明，否则编译遇到调用函数的语句时，系统就会认为被调用函数没有定义；如果把自定义函数的定义放在主函数前面，则自定义函数的声明可以省略。

各个自定义函数之间都是相互独立、相互平行的,没有排列上的先后顺序之分，因此自定义函数的声明顺序及其函数的定义排列顺序无关。

5.2.3 函数的调用

自定义函数在完成定义之后，就可以在程序中调用这个函数了。调用函数时，将实际参数传递给形式参数，并执行自定义函数的函数体语句，以实现函数相应的功能。实际参数和形式参数在数据类型、数量、顺序上必须完全一致，否则参数传递将会出现错误，导致出现语法错误或函数调用结果发生错误。

C语言规定，除了 main() 函数不能被调用以外，所有的标准库函数和自定义函数都是可以被调用的。调用其他函数的函

C language requires to first define function and then invoke it, if the definition of customized function is put behind the main function, then the declaration must be made for function before the main function, or else the system would consider the invoked function has not been defined when the compiler encountering the statement of invoking function; If the definition of customized function is put before the main function, then the declaration of customized function can be omitted.

Each customized is mutually independent and parallel, there is no sequence for ranking, therefore, the declaration sequence of customized function has no relation with the ranking sequence of function definition.

5.2.3 The Invocation of Function

After the customized function accomplishes the definition, this function can be invoked in program. When invoking function, transmitting the actual parameter to formal parameter, and performing the function body statement of customized function, so as to realize the corresponding function. The actual parameter and formal parameter must be consistent in data type, quantity and sequence, or else there will be error in transmitting data, resulting in the emergence of error in grammar or function invocation result.

As it is prescribed by C language, except for main() function can not be invoked, all of the standard library functions and customized functions can be invoked. The function utilized for invoking other

数称为"主调函数",被调用的函数称为"被调函数"。

调用没有返回值的函数时,只需要将函数作为一条语句进行处理即可,其调用的一般格式为

函数名(实参列表);

其中,实参是指调用该函数时所用到的全部实际参数,简称实参,实参列表中的实际参数在数据类型、数量、排列顺序上与形式参数完全一致,实参列表一般格式为

实参1,实参2,…,实参n

【例5.1】采用调用函数的方法,在屏幕上输出5行9个"*"的图案,如下所示。

```
*********
*********
*********
*********
*********
```

【解题思路】图形中每行的"*"都是相同个数,所以可以自定义一个函数prt(),输出一行10个"*"的图案,在主函数中用for循环实现调用5次prt()函数,即可输出显示规定图形。

调用有返回值的函数时,可以将该函数看成一个普通的变量进行使用,一般有

functions is called "main invoking function", the invoked function is called "invoked function".

While invoking function without return values, it is necessary to deal with the function as one statement, the general format for invocation is:

Function name (actual parameter list);

Among them, the actual parameter refers to all of the actual parameters utilized for invoking this function, the short name is actual parameter, the actual parameters in actual parameter list are consistent with formal parameters in data type, quantity and ranking sequence, the general format for actual parameter list is:

Actual parameter 1, actual parameter 2, …, actual parameter n

【Example 5.1】Applying the method for invoking function, outputting the patterns for five lines of nine "*" on the screen, the specific details are as following.

```
*********
*********
*********
*********
*********
```

【Problem-Solving Ideas】The "*" in each line of patterns has the same quantity, therefore, one function prt () can be customized, outputting the patterns with ten "*" in each line, applying for loop to realize the five times of invocations for prt () function in main function, namely, outputting to display the specified pattern.

When invoking function with return values, this function can be taken as the

```c
1  #include <stdio.h>
2  void prt()
3  {
4      int j;
5      for(j=0;j<10;j++)
6          printf("*");
7      printf("\n");
8  }
9  void main()
10 {
11     int k;
12     for(k=0;k<5;k++)
13         prt();
14 }
```

以下三种方式：

（1）把函数返回值赋给主调函数的某个变量，即

变量 = 函数名（实参列表）；

例如, c=max（a, b）;

该语句就是将主调函数中定义的变量 a 和 b 作为实际参数，分别传给 max() 函数中的变量 x 和 y，最后将 max 函数的返回值赋给主调函数中定义的变量 c。

（2）函数出现在一个表达式中参与运算，这种表达式称为函数表达式。

例如, if（a>=max（b, c））

该语句先计算 max（b, c）的值, 函数的返回值为 b、c 中的大数，再与变量 a 进行大小判断。

（3）函数调用作为一个函数的实参，其实质上也是函数表达式调用的一种。

例如, max（a, max（b, c））；

该语句先计算 max（b, c）的值, 再

ordinary variable for application, there are primarily the following three methods:

(1) Assigning the return value of function to certain variable of main invoking function, that is:

Variable=Function name (actual parameter list);

For example, c=max (a, b);

This statement is to take the variables a and b defined in main function as actual parameter, then transmitting to the variables x and y in max() function, at last, assigning the return value of max function to the variable c defined in main function.

(2) When the function appears in one expression to participate in operation, this expression is called function expression.

For example, if (a>=max (b, c))

This statement will first calculate the value of max (b, c), the return values of function are the greatest numbers in b and c, then comparing it with the variable a in size.

(3) Function invocation is a actual parameter for function, in nature, it is a kind of function expression invocation.

For example, max (a, max (b, c)) ;

This statement will first calculate the value of max (b, c), then taking the return

把函数的返回值，即 b、c 中的大数作为 max() 函数的一个实际参数，再次调用 max() 函数，最后的计算结果就是求得 a,b,c 三个数中的最大值。

【例 5.2】采用调用函数的方式，在屏幕上输出以下图形。（有参数无返回值的自定义函数的调用）

```
*
***
*****
*******
*********
```

【解题思路】图形中每行输出的"*"的个数分别是 1、3、5、7、9，第一行 1 个，以后每增加 1 行增加 2 个"*"，自定义一个函数 prt() 用来输出一行的"*"，每一行都用 for 循环实现不同数量的"*"，每行"*"的个数是可变的，将它设置成形式参数，在主函数中用 for 循环调用 5 次 prt() 函数，实际参数每次增量为 2。

value of function, which is the greatest number in b and c as the actual parameter for max() function, then invoking max() function again, at last, the calculation result is to obtain the greatest value among the three numbers of a, b and c.

【Example 5.2】Applying the method of function invocation, outputting the following pattern on screen. (The invocation of customized function with parameters and without return values)

```
*
***
*****
*******
*********
```

【Problem-Solving Ideas】The number of "*" output in each line of the pattern are 1, 3, 5, 7, and 9 respectively, there is one in the first line, after that, increasing two "*" in each added line, customizing one function prt () to output the "*" in each line, the "*" of different quantities will be realized in each line by for loop, the number of "*" in each line is changeable, setting it as formal parameters, invoking the prt function for five times in main function by for loop, the increment of actual parameter for each time is 2.

```c
#include <stdio.h>
void prt(int i)
{
    int j;
    for(j=0;j<i;j++)
        printf("*");
    printf("\n");
}
void main()
{
    int k;
    for(k=0;k<5;k++)
        prt(2*k+1);
}
```

Fifth Task: Categorizing and Summarizing Student Scores

【例5.3】用调用函数的方式，求两个数中的最大值，在主函数中输入两个数。

【解题思路】定义一个函数max()，实现将大的数放在变量x，小的数放在变量y，主函数中输入两个数和显示较大的数。

【Example 5.3】 By means of function invocation, calculating the maximum value in two numbers, inputting two numbers in main function.

【Problem-Solving Ideas】 Defining one function max(), realizing to put the greater number in variable x, the less number in variable y, inputting two numbers in main function and displaying the greater number.

```c
#include <stdio.h>
int max(int x,int y)
{
    int t;
    if(x<y)
    {
        t=x;
        x=y;
        y=t;
    }
    printf("max函数中x=%d, y=%d\n",x,y);
    return x;
}
void main()
{
    int x,y,m;
    printf("请输入两个整数：");
    scanf("%d%d",&x,&y);
    m=max(x,y);
    printf("调用max函数后x=%d,y=%d\n",x,y);
    printf("%d和%d中较大数是：%d",x,y,m);
}
```

注意：

（1）程序的执行总是从main()函数开始，在main()函数中遇到调用其他函数的语句，则转向其他函数执行，执行完毕后返回到main()函数，在main()函数中结束整个程序的运行。

（2）【例5.3】自定义函数max(int x, int y)中，int x和int y是形式参数，在该函数未被调用时没有确定的值，只是形式

Attention:

(1) The implementation of program always starts from main function, when encountering the statements that invoking other functions in main function, then turning to perform other functions, after implementation, returning to main function, ending the operation of entire program in main function.

(2) In customized function max(int x,int y) of 【Example 5.3】, int x and int y are both formal parameters, when this function is not invoked, there is no determined value except

上的参数，形参在函数调用时必须有相应的实际参数，main 函数中 "m=max(x,y);" 语句中的 x 和 y 就是函数调时的实际参数。

（3）在函数调用时，形参才被分配存储单元，接受实参传来的值，函数调用结束后形参释放存储单元，回收空间。因此实参和形参是不同的变量，可以使用相同的名字。函数调用时，将实参的值复制一份传递给形参，这种值的传递方式称为值传递，值传递是单向的，只能从实参传向形参，形参并不能返回到主调函数中改变实参。从【例5.3】运行后的结果中也可以看出值传递的方向，在 max 函数执行后，max 函数中的 x 变量保留的是 x 和 y 中的较大值，返回到主函数中时，x 变量依然是原本输入的数值，并没有发生变化。

1. 函数的嵌套调用

在某些情况下，在被调函数中又调用其他函数，称为函数的嵌套调用。

具有返回值的函数，在数值返回到主调函数时是一个普通数据类型的具体数值，因此它就可以作为其他有参函数的实际参数进行使用，此时就是函数的

for the formal parameter, formal parameter must have corresponding actual parameter in function invocation, in main function "m=max (x, y)"; The x and y in statement are the actual parameters for function invocation.

(3)As for function invocation, the formal parameter will be assigned to storage units to accept the value transmitted by actual parameter, when the function invocation ends, the formal parameter will release the storage units to recycle the space. Therefore, the actual parameter and formal parameter are different variables, the same names can be utilized for them. As for function invocation, duplicating one piece of the value of actual parameter and transmitting it to formal parameter, the transmitting method of such value is called value transmitting, value transmitting is unidirectional, which can only be transmitted to formal parameter, formal parameter can not return to main function to change the actual parameter. The transmitting direction of value can be observed from the operation result of【Example 5.3】, after the operation of max function, the variable x in max function will preserve the greater values in x and y, when returning to the main function, the x variable will still be the original input value, there is no change.

1. The Embedding and Invocation of Function

In some cases, other functions will be invoked in invoked functions, which is called the embedding invocation of function.

As for the function possessing return value, which is a specific value for ordinary data type when the values are returning to main function, therefore, it can be taken

Fifth Task: Categorizing and Summarizing Student Scores

嵌套调用。函数的嵌套调用示意图如图 5.1 所示。

as the actual parameter of other parameter functions for usage, at this moment, it is the embedding invocation of function. The function embedding invocation schematic diagram was presented in Figure 5.1.

图 5.1 函数嵌套调用
Figure 5.1 Function Embedding and Invocation

C 语言支持多层函数调用，无论函数在何处被调用，调用结束后，总是会回到调用该函数的地方。

【例 5.4】采用调用函数的方式，求三个数中的最大值，在主函数中输入三个数并输出最大值。

【解题思路】定义一个函数 max 求两个数中的较大值，第一次调用 max 函数求变量 a 和变量 b 中的大值，并作为第二次调用 max 函数的其中一个实际参数与变量 c 求较大值，最后返回的值即为变量 a、b、c 三个中的最大值。

C language supports multilevel function invocation, no matter the function is invoked in which place, after invocation, it will be invoked to the original place.

【Example 5.4】By means of function invocation, calculating the greatest value among the three numbers, inputting three numbers and outputting the greatest value in main function.

【Problem-Solving Ideas】Defining one function max to calculate the greater value of two numbers, for the first time, invoking the max function to obtain the greater value between the variable a and variable b, which will be taken as one of the actual parameters for invoking the max function for the second time, calculating the greater value in comparison with the variable c, at last, the returned value is the greatest value among the three variables of a, b and c.

```
1  #include <stdio.h>
2  int max(int x,int y)
3  {
4      int t;
5      if(x>y)
6          return x;
7      else
8          return y;
9  }
10 void main()
11 {
12     int a,b,c,m;
13     printf("请输入三个整数：");
14     scanf("%d%d%d",&a,&b,&c);
15     m=max(max(a,b),c);
16     printf("这三个数中最大的是：%d",m);
17 }
```

2. 函数的递归调用

一个函数在它的函数体内调用它自身，称为函数的递归调用，这种函数称为递归函数。

C语言允许函数的递归调用，在递归调用中，主调函数同时又是被调函数，执行递归函数将反复调用其本身，每调用一次就进入新的一层。为了防止递归调用无终止地进行，必须在函数内有终止递归调用的方式，常用附件条件判断，满足某种条件后就不再做递归调用，然后逐层返回。

因此编写递归函数有两个要点：确定递归公式和根据公式确定递归函数结束的条件。

【例5.5】利用递归调用的方式，编程

2. The Recursion and Invocation of Function

One function can invoke itself in the function body, in this case, it was called the recursion and invocation of function, and such function is called recursion function.

C language allows the recursion and invocation of functions, in the process of recursion and invocation, the main function for invoking is also the invoked function, performing the recursion function would invoke itself repeatedly, it will enter a new level once it is invoked. For the purpose of avoiding the endless recursion and invocation, there must be a method in function to end the recursion and invocation, judging by common attachment conditions, the recursion and invocation will not performed again once certain conditions are met, then returning level by level.

Therefore, there are mainly two focal points for compiling recursion function: Determining the recursion formula and deciding the ending conditions for recursion function in accordance with the formula.

【Example 5.5】By means of recursion

求 n！的值，在主函数中输入 n 的大小，并显示最后的结果。

【解题思路】求解 n 的阶乘可以分解为当 n=1 时，n=1；当 n>1 时，n！=n*(n-1)！。因此，可以自定义一个函数 jc(int i) 求 i 的阶乘，当 n>1 时函数 jc(n) 求 n！，函数 jc(n-1) 求 (n-1)！，当 n=1 时 jc(1)=1。在函数 jc() 的定义中又调用了函数 jc() 自身，这就是函数的递归调用，而 n=1 就是递归终止的条件，jc() 函数的递归调用就不会无终止地进行下去。

and invocation method, programming to calculate the value of n!, inputting the value of n in main function, and displaying the final result.

【Problem-Solving Ideas】The factorial for calculating n can be divided into when n=1, n=1; When n>1, n!=n*(n-1)!. Therefore, one function jc(int i) can be defined to calculate the factorial of i, when n>1, calculating n by function jc(n), calculating (n-1)! by function jc(n-1), when n=1, jc(1)=1. Invoking the function jc() itself in the definition of function jc(), this is the recursion and invocation of function, while n=1 is the recursion ending condition, the recursion and invocation of function jc() will not last constantly.

```c
#include <stdio.h>
long jc(int i);
void main()
{
    long s;
    int n;
    printf("请输入n的值：");
    scanf("%d",&n);
    s=jc(n);
    printf("%d! =%ld\n",n,s);
}
long jc(int i)
{
    long j;
    if(i==1)
        j=1;
    else
        j=i*jc(i-1);
}
```

5.2.4 数组做函数参数的应用

数组用作函数的参数有两种形式：一种是把单个的数组元素作为实参使用，

5.2.4 The Application of Array as Function Parameter

There are two methods for taking arrays as the parameters of function: First, utilizing the single array element as the actual parameter, second, taking array names as the

另一种是把数组名作为函数的形参和实参使用。

1. 单个数组元素作为实参

单个的数组元素用作函数的实参时，与其他类型普通变量做实参没有任何区别，在发生函数调用时，把单个数组元素的值传递给形参，实现单向的值传递，其调用方式与普通变量一样。

【例5.6】从键盘输入一串字符，编写一个函数，统计该字符串中数字的个数。

【解题思路】自定义一个函数count()，判断字符是否是数字，如果是数字则返回1，否则返回0。主函数中输入字符串存放在字符数组中，因字符串存储时是以"\0"结尾，因此以存储的字符不是"\0"作为循环条件来构造一个循环，以字符数组的每个元素作为实际参数来调用count函数，对其返回值进行累计，即可统计该字符串中数字的个数，最后在主函数中输出统计出的个数。

此例中，字符数组c中的单个元素c[i]，跟普通变量一样，作为函数的实参来调用count函数，将其存储的字符传递给count函数的变量i参与运算，主函

formal parameters and actual parameters of functions.

1. Taking Single Array Element as Actual Parameter

When taking single array element as the actual parameters for functions, there is no difference for taking ordinary variables of other types as the actual parameters, when invoking the function, transmitting the value of single array element to formal parameter to realize the unidirectional value transmitting, its invocation method is the same as that of ordinary variables.

【Example 5.6】Inputting one string of characters from keyboard, compiling one function, calculating the numeric scale in this character string.

【Problem-Solving Ideas】Customizing one function count(), judging if the characters are numbers, if they are numbers, then returning to 1, or else returning to 0. Inputting the character strings in main function and storing them in character arrays, because the character string is ended with "\0" while storing, therefore, constructing a loop based on the loop condition that the storage characters are not "\0", taking each element in character arrays as the actual parameter to invoke the count function, accumulating its returned values, then the number scale can be calculated in this character strings, at last, outputting the calculated number in main function.

In this example, the single element c[i] in character array c is the same as ordinary variable, it can be taken as the actual parameter of function to invoke the count function, transmitting its storage characters to the variable i of count function to participate in

Fifth Task: Categorizing and Summarizing Student Scores

```c
#include <stdio.h>
int count(char i)
{
    if(i>='0'&&i<='9')
        return 1;
    else
        return 0;
}
void main()
{
    int i,s=0;
    char c[30];
    printf("请输入一串字符：");
    scanf("%s",c);
    while(c[i]!='\0')
    {
        s=s+count(c[i]);
        i++;
    }
    printf("这串字符有 %d 个数字。",s);
}
```

数中的字符数组 c 本身并不会发生任何变化。

2. 数组名作为函数的形参和实参

数组名作为函数参数时，即可以是形参，也可以是实参，要求形参和对应的实参都必须是相同数据类型的数组，并且都必须有明确的数组定义。

数组名不但代表的是数组元素的共同名字，而且还代表着数组中第一个元素存放的首地址，所以数组名作为函数参数时，实参是将数组元素的首地址传递给形参，这种传递方式称为地址传递。因为形参数组和实参数组是共用一段相同的数据内存空间，因而在调用函数的过程中，形参数组元素值的改变，也就

the operation, no change will be found in the character array c itself in main function.

2. Taking the Array Name as the Formal Parameter and Actual Parameter of Functions

When taking the array name as the function parameter, it can be formal parameter or actual parameter, it is required that the formal parameter and the corresponded actual parameter must be the arrays of same data type, and they must have the specific array definition.

Array name not only represents the common names of array elements, but also represents the primary address for storing the first element in arrays, therefore, when taking array name as the function parameters, the actual parameter can transmit the initial address of array element to formal parameter, such transmitting method is called address transmitting. Because the formal parameter arrays and the actual parameter

使得实参数组元素值做了相同的改变，因此地址传递可以认为是双向传递，即实参数组将地址传递给形参数组，形参数组各元素在调用函数过程中发生的变化也保存到了实参数组各元素中。

【例5.7】某一篮球队原本有10名队员，后因个人原因有一人退队，编程实现10名队员名单的更新，要求10名队员的名字和退出球队的队员的名字都从键盘输入。

【解题思路】自定义一个函数del()实现数组元素的删除功能，需要删除的数组元素用查找的方式进行确定，在主函数中定义一个二维字符数组存储10人姓名，输入10名球队队员的名字，再输入退队队员的名字，作为其中一个实际参数来调用函数del()，实现删除队员名字的功能，最后输出更新后的球队队员名单。

arrays have the common data storage space, therefore, in the process of invoking the functions, the changes made on the element value of formal parameter array also happen on the element value of actual parameter array, hence, address transmitting can be taken as the bidirectional transmitting, that is to transmit the actual parameter array to formal parameter array, the various elements of formal parameter arrays have changed in the process of invoking function, which have been stored in the various elements of actual parameter arrays.

【Example 5.7】There are ten team members in certain basketball team originally, due to personal reasons, one member has quitted the team, programming can realize the renovation of lists for ten team members, it is required that the names of ten team members as well as the names for the members who have quitted the team can be input from keyboard.

【Problem-Solving Ideas】Customizing one function del() to realize the deletion function of array elements, the array elements need to be deleted can be determined by searching method, defining a two-dimensional character arrays in main function to store the names of ten people, inputting the names of ten team members, and then inputting the names of members who have quitted the teas, invoking the function del() as one of the actual parameters, realizing the function for deleting the names of team members, at last, outputting the renovated team member lists.

Fifth Task: Categorizing and Summarizing Student Scores

```c
#include <stdio.h>
#include <string.h>
#define N 10
void del(char name[N][20],char dname[20])
{
    int i,j;
    for(i=0;i<N;i++)
    if(strcmp(name[i],dname)==0)
    {
        break;
    }
    for(j=i;j<N-1;j++)
    {
        strcpy(name[j],name[j+1]);
    }
}

void main()
{
    int i;
    char name[N][20],t[20];
    printf("请输入%d个队员的名字：\n",N);
    for(i=0;i<N;i++)
        scanf("%s",name[i]);
    printf("请输入退队队员的名字：\n");
    scanf("%s",t);
    del(name,t);
    printf("新队员名单\n");
    for(i=0;i<N-1;i++)
        printf("%s\n",name[i]);
}
```

此例中，主函数中的 name 数组和 t 数组都是实参数组，del() 函数中的 name 数组和 dname 数组都是形参数组，调用 del() 函数时以数组名作为实参，是将主函数中 name 数组和 t 数组的首个元素的首地址传递给 del() 函数中的 name 数组和 dname 数组，两个 name 数组共用一段相同的存储空间，t 数组和 dname 数组共用一段相同的存储空间，所以对 del 函数中的 name 数组和 dname 数组的元素进行改变的同时，其实就是对主函数中 name 数

In this example, the name array and t array in this main function are both actual parameter arrays, the name array and dname array in del function are both formal parameter array, taking array name as actual parameter when invoking del() function, it is to transmit the initial address of initial element in name array and t array of del() function to the name array and dname array of del function, the two name arrays have the common storage space, t array and dname array possess the common storage space, therefore, when changing the elements of name array and dname array in

组和 t 数组的元素的改变。比如甲、乙二人同住一间宿舍，对甲的宿舍进行重新装修，那么实际上乙的宿舍也被同样的重新装修了。这就是地址传递方式中，形参数组和实参数组的双向传递效果。

注意：

（1）数组名作为函数的参数，必须在主调函数和被调函数中分别进行定义，且数据类型必须一致，否则会出现错误。

（2）C 语言编译系统对形参数组大小不做检查，所以形参数组可以不指定大小。但如果一定要指定形参数组的大小，那么其长度必须小于实参数组的长度，否则会因为形参数组的部分元素没有确定值而导致计算结果错误。

5.3 任务实现

【任务要求】某班有 40 位学生参加了期终考试，考了三门课，分别是 C 语言、高等数学、英语，40 位学生分成 5 个组，每个组的学生人数可以不同，老师想统计以下信息：①统计小组一门课程的总分及平均分；②统计小组若干门课程的

del function, it is to change the elements of name array and t array in main function. For example, A and B live in the same dormitory, redecorating the dormitory of A means to redecorate the dormitory of B. This is the bidirectional transmitting result for formal parameter array and actual parameter array in address transmitting method.

Attention:

(1) As for taking the array name as the parameter of function, it is necessary to define it in the main invocation function and invoked function respectively, and the data type must be consistent, or else there will be error.

(2) The C language compiling system would not check the size of formal parameter array, hence, the size of formal parameter array does not need to be designated. However, if the size of formal parameter array must be designated, then its length must be less than the length of actual parameter array, or else the calculation result error will be caused due to the undetermined value for some elements in the formal parameter array.

5.3 Task Implementation

【Task Requirements】There are forty students have participated in the final examination in certain class, three subjects have been examined, they are C language, superior math, and English respectively, the forty students have been divided into five groups, the number of students in each group does not need to be the same, the teacher wants to calculate the following

Fifth Task: Categorizing and Summarizing Student Scores

总分及平均分；③输出排序后小组三门课成绩单。

【任务分析】本任务要求的功能较多，为了使程序的结构简明清晰，将此任务分解成多个函数模块进行，主函数 main 制作菜单并根据需要调用相应的函数，A 函数统计小组一门课程的总分及平均分，B 函数统计小组若干门课程的总分及平均分，C 函数对小组成绩进行排序后输出。

5.3.1 统计小组一门课程的总分和平均分

定义函数 sumf() 进行输入本小组所有同学的分数，并求本小组的总分，主函数制作菜单并根据需求调用相应函数，在主函数中输入本小组学生的个数，调用 sumf() 函数求总分并求平均分，最后输出总分和平均分。

information: ① Calculating the total scores and average scores of one subject for groups; ② Calculating the total scores and average scores of several subjects for groups; ③ Outputting the transcripts of three subjects for groups after ranking.

【Task Analysis】There are more functions required by this task, for making the structure of program more clear and simple, decomposing this task into several function modules to perform, the main function main would make the menu and invoke the corresponding function in accordance with the requirements, A function would calculate the total scores and average scores of one subject for groups, B function would calculate the total scores and average scores of several subjects for groups, C function would rank the group transcripts and output.

5.3.1 Calculating the Total and Average Scores of One Subject for Groups

Defining function sumf() and input the scores of all students in this group, and calculating the total scores of this group, the main function would make menu and invoke the corresponding function in accordance with the requirements, inputting the number of students of this group in main function, invoking the sumf() function to calculate the total scores and average scores, at last, outputting the total scores and average scores.

```c
1   #include "stdio.h"
2   float sumf(int n)
3   {
4       int x,i;
5       float s=0;
6       printf("请输入本小组的考试成绩: ");
7       for(i=0;i<n;i++)
8       {
9           scanf("%d",&x);
10          s+=x;
11      }
12      return s;
13  }
14  void main()
15  {
16      int k,n;
17      float sum,avg;
18      char ch;
19      do
20      {
21          printf("*****************************************\n");
22          printf("\t   班级成绩统计\n");
23          printf("*****************************************\n");
24          printf("1、统计小组一门课程的总分及平均分\n",n);
25          printf("2、统计小组若干门课程的总分及平均分\n");
26          printf("3、输出小组排序后三门课程的成绩单\n");
27          printf("4、退出\n");
28          printf("\n请输入1~4进行选择: ");
29          scanf("%d",&k);
30          if(k==1)
31          {
32              printf("请输入统计的小组的人数 n = ");
33              scanf("%d",&n);
34              sum=sumf(n);
35              avg=sum/n;
36              printf("本小组的总分=%.0f\t平均分=%.1f\n\n",sum,avg);
37          }
38      }while(k!=4);
39  }
```

5.3.2 统计小组若干门课程总分和平均分

定义函数 avgevery() 进行输入本小组所有同学若干门课程的分数，并求本小组本课程的总分和平均分。主函数制作菜单并根据需求调用相应函数，在主函数中输入本小组学生的个数和课程门数，最后调用 avgevery() 函数求各门课程的总分和平均分，并进行输出。

5.3.2 Calculating the Total and Average Scores of Several Subjects for Groups

Defining the function avgevery() to input the scores for several subjects of all the students in this group, calculating the total and average scores of this subject for this group. The main function would make menu and invoke the corresponding function in accordance with the requirements, inputting the number of students and their subject scores in main function, at last, invoking the avgevery () function to calculate the total and average scores of each subject for students and output them.

Fifth Task: Categorizing and Summarizing Student Scores

```c
#include "stdio.h"
void avgevery(int n,int km)
{
    int x,i,j;
    float s,avg;
    for(j=1;j<=km;j++)
    {
        s=0;
        printf("\n请输入本小组第%d门考试成绩: ",j);
        for(i=1;i<=n;i++)
        {
            scanf("%d",&x);
            s+=x;
        }
        avg=s/n;
        printf("第%d课程的总分=%.0f\t平均分=%.1f\n",j,s,avg);
        printf("\n");
    }
}

void main()
{
    int k,n,km;
    float sum,average;
    do
    {
        printf("\n*****************************************\n");
        printf("\t   班级成绩统计\n");
        printf("*****************************************\n");
        printf("1、统计小组一门课程的总分及平均分\n",n);
        printf("2、统计小组若干门课程的总分及平均分\n");
        printf("3、输出小组排序后三门课程的成绩单\n");
        printf("4、退出\n");
        printf("\n请输入1~4进行选择: ");
        scanf("%d",&k);
        if(k==2)
        {
            printf("请输入统计的小组的人数 n = ");
            scanf("%d",&n);
            printf("请输入要统计的课程门数 km = ");
            scanf("%d",&km);
            avgevery(n,km);
        }
    }while(k!=4);
}
```

5.3.3 输出排序后小组的成绩单

为了方便运行调试，假设班上只有 5 名同学。定义一个 input() 函数，用于输入每个同学的姓名和三门课的成绩；定义一个 sumavg() 函数，用于求每个同学的总分和平均分；定义一个 px() 函数，用于根据总分的高低进行降序排序，交换数据时，该名同学的姓名、三门课成绩、总

5.3.3 Outputting the Transcripts of Groups after Ranking

For the convenience of operating invocation, supposing that there are only five students in the class. Defining a input() function and utilizing it to input the name and scores for three subjects of each student; defining a sumavg() function, utilizing it to calculate the total and average scores of each student; defining a px() function, utilizing it to rank the total scores by descending

分、平均分等全部都要进行交换；定义一个print()函数，用于成绩单的输出，包括序号、学生的姓名、三门课成绩、总分、平均分；主函数main()定义几个二维数组用于存放学生的姓名、三门课成绩，定义两个一维数组存放每名同学的总分和平均分，最后依次调用输入函数input()、求总分及平均分函数sumavg()、排序函数px()、输出函数print()，完成统计小组学习成绩的功能。

order, when changing data, the name, scores for three subjects, total scores and average scores of this student would be changed; defining a print() function, utilizing it to output transcripts, including serial number, student name, the scores for three subjects, total scores and average scores; The main function main can define several two-dimensional arrays to store the student names and their scores for three subjects, defining two one-dimensional array to store the total score and average score of each student, at last, invoking the input() function input, the function sumavg() that calculating the total scores and the average scores, the ranking function px() and the output function print() sequentially, accomplishing the function for calculating the learning scores of group students.

```c
#include "stdio.h"
#include "string.h"
#define N 5
void input(int score[N][3],char name[N][10])
{
    int i,j;
    for (i=0;i<N;i++)
    {
        printf("第%d个同学的姓名及三门课的成绩：",i+1);
        scanf("%s",name[i]);
        for(j=0;j<3;j++)
            scanf("%d",&score[i][j]);
    }
}

void sumavg(int score[N][3],float sum[],float avg[])
{
    int i,j;
    for(i=0;i<N;i++)
    {
        for(j=0;j<3;j++)
            sum[i]=sum[i]+score[i][j];
        avg[i]=sum[i]/3.0;
    }
}
```

Fifth Task: Categorizing and Summarizing Student Scores

```c
25  void px(int score[][3],float sum[],float avg[],char name[][10])
26  {
27      int i,j;
28      float t;
29      char nn[10];
30      for(i=0;i<N-1;i++)
31          for(j=0;j<N-1-i;j++)
32              if(sum[j]<sum[j+1])
33              {
34                  t=sum[j];sum[j]=sum[j+1];sum[j+1]=t;
35                  t=avg[j];avg[j]=avg[j+1];avg[j+1]=t;
36                  t=score[j][0];score[j][0]=score[j+1][0];score[j+1][0]=t;
37                  t=score[j][1];score[j][1]=score[j+1][1];score[j+1][1]=t;
38                  t=score[j][2];score[j][2]=score[j+1][2];score[j+1][2]=t;
39                  strcpy(nn,name[j]);
40                  strcpy(name[j],name[j+1]);
41                  strcpy(name[j+1],nn);
42              }
43  }


44  void print(int score[ ][3],float sumr[ ],float avgr[ ],char name[ ][10])
45  {
46      int i,j;
47      printf("**************************************************\n");
48      printf("\t\t\t成绩单\n");
49      printf("**************************************************\n");
50      printf("序号\t姓名\t课1\t课2\t课3\t总分\t平均分\n");
51      for(i=0;i<N;i++)
52      {
53          printf("%d:\t",i+1);
54          printf("%s\t",name[i]);
55          for(j=0;j<3;j++)
56              printf("%d\t",score[i][j]);
57          printf("%.0f\t%.1f\t",sumr[i],avgr[i]);
58          printf("\n");
59      }
60  }


61  void main()
62  {
63      int score[N][3];
64      char name[N][10],nn[10];
65      float sumr[N]={0},avgr[N];
66      input(score,name);
67      sumavg(score,sumr,avgr);
68      px(score,sumr,avgr,name);
69      print(score,sumr,avgr,name);
70  }
```

【练习与提高】

1. 编写两个函数，分别求两个整数的最大公约数和最小公倍数，用主函数输入两个函数、调用两个函数，并输出结果。

2. 已知一元二次方程为 $ax^2+bx+c=0$，编写3个函数，分别求当 $\Delta=b^2-4ac$ 的值大于0、等于0和小于0时的根，在主函数中输入 a、b、c 的值，并输出方程的根。

3. 编写一个判断素数的函数，主函数中输入数据区间（如输入两个数1和100代表1~100之间），在屏幕上显示出该数据区间内所有的素数，要求每行只显示5个数。

4. 编写几个函数，实现以下功能：

（1）输入10个职工的姓名和工号。

（2）按职工号的大小排序，显示所有职工名单。

（3）要求输入一个职工号，查找出该职工的姓名，从主函数中输入要查找的职工号，输出该职工姓名。

【Practice and Improvement】

1. Compiling two functions, calculating the greatest common divisor and lowest common multiple of two integers respectively, utilizing the main function to input two functions, invoking the two functions, and outputting the result.

2. It is known that the quadratic equation with one unknown is $ax^2+bx+c=0$, compiling three functions, by means of quadratic formula to calculate the value when $\Delta=b^2-4ac$ is greater than 0, equal to and less than 0 respectively, inputting the values of a, b and c in main function, and outputting the equation root.

3. Compiling a function to judge prime number, inputting the data section in main function (if inputting the two numbers of 1 and 100, signifying that it is between 1 to 100), displaying all of the prime numbers in this data section on the screen, it is required that only five numbers can be displayed in each line.

4. Compiling several functions to realize the following functions:

(1) Inputting the names and job number of ten personnel.

(2) Ranking according to the job number by descending order, displaying all of the personnel lists.

(3) It is required to input one personnel number, finding out the name of this personnel, inputting the searching job number in main function, outputting the name of this personnel.

任务 6
学生成绩单制作

Sixth Task: Producing Students Report Cards

【知识目标】
1. 熟练掌握函数的声明、定义和调用。
2. 能编写和调用无参数函数、有参函数。
3. 掌握函数的嵌套调用和递归调用。

【能力目标】
1. 能够熟练应用函数解决实际问题。
2. 学会使用局部变量和全局变量。

【重点、难点】
1. 有参函数形参和实参的值传递方式。
2. 函数的嵌套。

【课程思政】
1. 通过结构化的程序分析，培养学生工程项目分析能力和管理能力，同时加强学生的团队精神和团队合作能力。
2. 通过递归函数的定义，强调言传身教的重要性。

【推荐教学方法】
通过教学做一体化教学，结合生活中常见的事例，使学生掌握知识点，学会编制程序流程图并进行程序的编写。

【推荐学习方法】
通过完成任务，在做中学、学中做，掌握实际技能与相关知识点。

【Knowledge Objective】
1. Proficient in the declaration, definition and invocation of function.
2. Capable of compiling and editing the parameterless function and parameter function.
3. Mastering the embedding invocation and recursion invocation of functions.

【Competency Objective】
1. Capable of applying the function to solve actual problems proficiently.
2. Learning to utilize the local variables and global variables.

【Focal and Difficult Points】
　　1. The transmitting method for the values of formal parameter and actual parameter of parameter function.
　　2. The embedding of function.

【Curriculum Ideological and Political Education】
　　1. By means of structural program analysis, cultivating the project program analysis capabilities and management capabilities of students, in the meantime, reinforcing the team spirit and the team cooperation capabilities of students.
　　2. By means of the recursion function definition, stressing the importance of teaching by personal example as well as verbal instruction.

【Recommended Teaching Method】
　　By means of integrated teaching, combining the common cases in daily life, making the students grasp knowledge points, learning to compile program flow diagram and edit programs.

【Recommended Learning Method】
　　By means of accomplishing tasks, learning by doing, doing by learning, mastering the actual technologies and relevant knowledge points.

6.1 任务描述

设计出完整的学生成绩单，增加每个同学的基本信息，包括学号、姓名、年龄、性别及每门课程的成绩、总分和平均分并进行排序。现在对学生成绩管理系统进行如下管理：①录入和显示每个同学的基本信息和每门课程的成绩；②计算每位同学的总分和平均分并进行排序。学生的姓名为字符型，学号为整型或者字符型；性别为字符型，年龄和每门课的成绩为整型，总分和平均分为整型或实数型。显然不能用一个数组来存放这一组数据。如何解决这个问题？虽然可以用多个数组把信息都储存起来，但是不同的数组之间建立联系是非常困难的，对数组分别赋值时，容易发生错位，

6.1 Task Description

Designing the complete student transcripts, increasing the basic information of each student, including student number, student name, age, gender and the scores for each subject, the total scores and average scores, then ranking them accordingly. At present, managing the student score management system by the following methods: ①inputting and displaying the basic information and scores for each subject of each student; ②calculating the total scores and average scores of each student and ranking them accordingly. The student name is of character type, the student number is of integer type of character type; The gender is of character type, the age and the scores for each subject are of integer type, the total scores and average scores are of integer type or real number type. Clearly, this set of data

分配内存不集中，不容易管理，所以在时间编程中很少使用这种方式。为了解决这个问题，C 语言提供了一种构造数据类型——结构体。

6.2 相关知识

结构体是 C 语言中另一种用户自定义的可用的数据类型，它允许存储不同类型的数据项。结构体本质上还是一种数据类型，但它可以包括若干个"成员"，每个成员的类型可以相同也可以不同，也可以是基本数据类型或者又是一个构造类型。

6.2.1 结构体类型的定义

结构体 (structure) 是由不同数据类型的数据组成的。组成结构体的每个数据称为该结构体类型的成员项，简称成员 (member)。在程序中使用结构体时，首先要对结构体的组成进行描述，即进行结构体类型定义，然后定义结构体类型的变量。

定义一个结构体类型的一般格式为

can not be stored by one array. How to solve this problem? Though the information can be stored by means of several arrays, it is very difficult to establish relations among different arrays, when assigning the values to arrays, there will be dislocation, it is not intensive to assign storage, which becomes difficult to manage, therefore, this method is seldom utilized in time programming. For solving this problem, C language provides a kind of construction data type——structure body.

6.2 Relevant Knowledge

Structure body is another kind of available data type customized by users in C language, it allows to store the data items of different types. In nature, the structure body is a kind of data type, which including several "members", the type of each member can be the same of different, which can be the basic data type or another construction type.

6.2.1 The Definition of Structure Body Type

The structure body is composed of data with different types. Each data that composing the structure body can be called the member item of this structure type, member for short. When utilizing the structure body in program, first of all, describing the composition of structure body, that is to define the structure body type, then defining the variables of structure body type.

The general format to define one structure type is:

```
struct 结构体类型名
{
    数据类型 成员名1;
    数据类型 成员名2;
    ... ...
    数据类型 成员名n;
}; //注意最后要有分号
```

其中,struct 是关键字,作为定义结构体类型的标志,后面紧跟的是结构体名,由用户自定义,花括号内是结构体的成员说明表,用来说明该结构体有哪些成员及它们的数据类型。花括号外的分号不能省略,它表示一种结构体类型说明的终止。

例如,定义一个表示日期的结构体类型。

```
struct date
{
    int year;    //年
    int month;   //月
    int day;     //日
};
```

【例 6.1】本任务中学生的基本信息见表 6.1,用结构体类型来实现。

```
Struct Structure body type name
{
    Data type  Member name 1;
    Data type  Member name 2;
    ... ...
    Data type  Member name n;
}; //Pay attention that there must be semicolon in the end
```

Among the, struct is the keyword, as a mark to define the structure body type, it is followed by the structure body name closely, which is customized by users, the member description table is presented in braces, signifying the members and their data types in this structure body. The semicolon beyond braces cannot be omitted, it signifies the ending of one structure body type.

For example, defining a structure body type to present data.

```
struct date
{
    int year;    //Year
    int month;   //Month
    int day;     //Day
};
```

【Example 6.1】The basic information of students in this task was presented in Table 6.1, which can be realized by structure body type.

表 6.1 学 生 基 本 信 息 表

学号	姓名	性别	功课1	功课2	功课3	平均分
id	name	sex	m1	m2	m3	avg
01	李小明	男	89	98	78	88.3

Table 6.1 Student Basic Information Table

Student number	Name	Gender	Subject 1	Subject 2	Subject 3	Average Score
id	name	sex	m1	m2	m3	avg
01	Li Xiaoming	male	89	98	78	88.3

The basic information structure body type of students was defined by the following statements.

```
struct stu
{
    char id[4];
    char name[10];
    char sex[4];
    int m1,m2,m3;
    float avg;
};
```

Defining a structure body type called stu, its members including id, name, sex, m1, m2, m3 and avg, they are the data items of different types.

6.2.2 The Definition of Structure Body Variables

There are the following three methods for defining the structure body variables.

1. Firstly, Declaring the Structure Body Type, Then Defining the Structure Body Variable.

For example

```
struct Student
{
    int stu_ID; //Student number
    char name[20]; //Name
    char sex; //Gender
    float score; //Scores
};
struct Student stu1 ,stu2 ;
```

Defining the two variables of stu1 and stu2 are both of struct Student structure body type, possessing the structure of struct Student type, as it was presented in Figure 6.1. By means of this method, declaring the separation of type and definition, the variables can be defined at any time after declaring the type, which is of flexible

1001	Zhang Ming	M	89.5
1002	Li Hua	F	98.0

图 6.1 结构体变量示意图
Figure 6.1 Structure Body Variable Schematic Diagram

活。在编写大型程序时，常采用此方式定义结构体变量。

2. 在声明结构体类型的同时定义结构体变量

例如：

```
struct Student
{
    int stu_ID; // 学号
    char name[20]; // 姓名
    char sex; // 性别
    float score ; // 成绩
}stu1 ,stu2;
```

该方式定义的一般形式如下：

```
struct 结构体名
{
    成员列表
}变量名列表；
```

这种方式中，声明类型和定义变量一起进行，能直接看到结构体的结构，较为直观，在遍写小程序时常用此方法。

3. 不指定结构体名而直接定义结构体变量

例如：

```
struct
```

application. When compiling the large-scale program, this method is utilized usually to define the structure body variables.

2.Defining the Structure Body Variable When Declaring the Structure Body Type

For example

```
struct Student
{
    int stu_ID; //Student number
    char name[20]; //Name
    char sex; //Gender
    float score ; //Scores
}stu1 ,stu2;
```

The general method for this defining method is:

```
struct Structure body name
{
    member list
}Variable name list;
```

By means of this method, the declaration of type and defining of variables can be performed simultaneously, the structure of structure body can be observed directly, which is more visual, such method has been utilized usually in compiling the little programs.

3.Don't Designate the Structure Body Name But to Define the Structure Body Variable Directly

For example

```
struct
```

Sixth Task: Producing Students Report Cards

```
    {
        int stu_ID; //Student number
        char name[20]; //Name
        char sex; //Gender
        float score ; //Scores
    }stu1 ,stu2 ;
```

The general format for this defining method is as following:

```
struct
{
    member list
}Variable name list;
```

Namely, don't display the structure body name directly, but to display the structure body variable directly. By means of this method, there is no structure body name, therefore, such structure body type cannot be utilized for defining other variables, there are less actual applications.

Attention:

(1) The members in structure body can also be the variables of a structure body type, as it was presented in Figure 6.2.

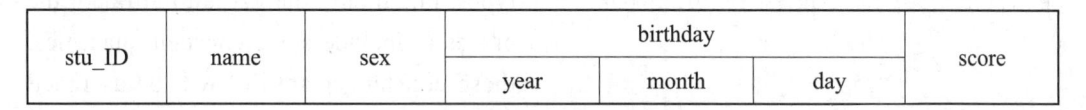

stu_ID	name	sex	birthday			score
			year	month	day	

Figure 6.2   The Data Structure Schematic Diagram of Structure Body

The following structure body can be presented according to Figure 6.2:

```
struct Date
{
    int year;
    int month;
    int day;
};
struct Student
{
    int stu_ID;
```

```
    char name[20];
    char sex;
    struct Date birthday; //birthday 为 struct Date 类型
    float score; //Scores
};
struct Student stu1,stu2;
```

首先声明一个 struct Date 类型，由 year、month、day 这 3 个成员组成；然后再声明 struct Student 类型，将其中的成员 birthday 指定为 struct Date 类型；最后定义 struct Student 类型的两个变量 stu1 和 stu2。

（2）结构体中的成员名可与程序中其他变量同名，但两者代表不同的对象，互不干扰。

6.2.3 结构体变量的初始化

和其他类型的变量一样，结构体变量可以在定义时进行初始化赋值，初始化列表是用花括号括起来的一些常量，这些常量依次赋给结构体变量中的成员。例如：

```
struct Student
{
int stu_ID;
char name[20];
char sex;
float score;
};
struct Student stu1 = {1001, "Zhang ping", 'M' ,78.5};
```

6.2.4 结构体变量的引用

在定义结构体变量以后，便可引用该

```
    char name[20];
    char sex;
    struct Date birthday; //birthday 为 struct Date Type
    float score; //Scores
};
struct Student stu1,stu2;
```

First of all, declaring a struct Date type, which is composed of three members of year, month and day; Then declaring the type of struct Student, designating the member birthday among them as the struct Date type; At last, defining the two variables of stu1 and stu2 of struct Student type.

(2)The member names in structure body can be the same as the variable names in other programs, but these two signifying different objects, which are non-interfering.

6.2.3 The Initialization of Structure Body Variable

The structure body variable can be assigned the initialized value in definition, which is the same as variables of other types, initializing the list is to utilizing the braces to include some constant quantities, these constant quantities will be assigned to the members in structure body variables sequentially. For example

```
struct Student
{
int stu_ID;
char name[20];
char sex;
float score;
};
struct Student stu1 = {1001, "Zhang ping", 'M' ,78.5};
```

6.2.4 The Citation of Structure Body Variable

After defining the structure body

变量。引用结构体变量成员的一般形式为

结构体变量名 . 成员名

例如：

stu1.stu_ ID 即第一名学生的学号

stu2. sex 即第二名学生的性别

"."是成员(分量)运算符，它在所有的运算符中优先级最高，因此可以把 stu1.stu_ID 作为一个整体看待。

如果成员本身又是一个结构体类型，则必须逐级找到最低级的成员才能使用。例如：

stu1.birthday.month 为第一名学生出生的月份。

x.m1=78

scanf("%s",&x.id); // 输入一个字符串送给结构体成员 x.id

printf("%s",x.id);

大家可以思考一下：

scanf("%s%s%d%d%d%d",&x); 能整体读入结构体变量的值吗？

printf("%s\t%s\t%5d%5d%5d\n",x); 能整体输出结构体变量的值吗？

答案显然是不能，它们的正确表达式为 scanf("%s%s%d%d%d",x.id,x.name,&x.m1,&x.m2,&x.m3); printf("%s\t%s\t%5d%5d%5d\n",x.id,x.name,x.m1,x.m2,x.m3);

variable, then citing this variable. The general method for citing the structure body variable member is

Structure body variable name. Member name

For example

stu1.stu_ ID is the student number for the first student

stu2. sex is the gender for the second student

"." is the member (weight) operator, it has the highest superiority in all of the operators, therefore, the stu1.stu_ID can be taken as an integrity.

If the member itself is a structure body type again, then it is necessary to find the inferior member level by level for application. For example:

stu1.birthday.month is the birth month for the first student.

x.m1=78

scanf("%s",&x.id); // Inputting one character string and send it to structure body members x.id

printf("%s",x.id);

Please reflect on it:

scanf("%s%s%d%d%d%d",&x); Can the value of structure body variable be read integrally?

printf("%s\t%s\t%5d%5d%5d\n",x); Can the value of structure body variable be output integrally?

The answer is clearly no, their correct expressions are:

scanf("%s%s%d%d%d",x.id,x.name,&x.m1,&x.m2,&x.m3); printf("%s\t%s\t%5d%5d%5d\n",x.id,x.name,x.m1,x.m2,x.m3);

【例 6.2】用键盘输入某学生的信息（包含学号、姓名、三门课的成绩）并在显示器上输出。

【解题思路】在定义一个结构体类型的同时定义了一个结构体变量 x，用 scanf() 函数完成结构体变量的输入，然后利用 printf() 函数进行输出。

【Example 6.2】Inputting the information of certain student by keyboard (including student number, student name and the scores of three subjects), outputting them on the display.

【Problem-Solving Ideas】While defining a structure body type, defining one structure body variable x, utilizing scanf() function to accomplish the inputting of structure body variables, then taking advantage of the printf() function to output.

```c
#include "stdio.h"
void main()
{
    struct
    {
        char id[6],name[10];
        int m1,m2,m3;
        float avg;
    } x;
    printf("请输入学生的信息（学号、姓名、三门课成绩）\n");
    scanf("%s%s%5d%5d%5d",x.id,x.name,&x.m1,&x.m2,&x.m3);
    printf("学生的信息为:\n");
    printf("%s\t%s\t%5d%5d%5d\n",x.id,x.name,x.m1,x.m2,x.m3);
}
```

【例 6.3】结构体变量的初始化和引用。

【解题思路】先定义一个结构体类型 struct Student，在定义结构体变量并赋值，

【Example 6.3】The initialization and application of structure body variables.

【Problem-Solving Ideas】First of all, defining one structure body type struct Student, next, defining structure body variable and assigning the values, then taking advantage

```c
#include <stdio.h>
#include <string.h>
struct Student //声明结构体类型
{
    int stu_ID; //学号
    char name[20]; //姓名
    float score; //成绩
};
void main()
{
    struct Student stu1 ={1001,"Sun Li",75.0}; //定义 stu1 变量并初始化
    struct Student stu2,stu3; //定义 stu2、stu3 变量
    stu2.stu_ID =1002; //引用结构体变量成员，并赋值
    strcpy(stu2.name,"Zhou Pi"); stu2.score =80.0;
    stu3 = stu1 ; //结构体变量相互赋值
    printf("学号\t 姓名\t成绩\n");
    printf("%d\t%s\t%3.1f\n",stu1.stu_ID,stu1.name,stu1.score);
    printf("%d\t%s\t%3.1f\n",stu2.stu_ID,stu2.name,stu2.score);
    printf("%d\t%s\t%3.1f\n",stu3.stu_ID,stu3.name,stu3.score);
}
```

of printf() function to output.

Attention:

(1) Various operations can be performed on structure body variable members and ordinary variables. For example

stu3.score = stu2. score +10;

sum = stu1.score + stu2.score + stu3.score ;

(2) It is available to cite the address of structure body variables. For example

```
scanf( "%f" ,&stu1,score);   //Inputting the value of stu1.
                                score
scanf( "%s" ,stu1.name);    //Inputting the character string
                                of stu1.name, pay attention not
                                to obtain address symbols&
```

6.2.5　The Definition of Structure Body Array

The relevant data regarding one student can be stored in one structure body variable, if there is a need to preserve and cope with the data of several students, clearly, it will come to utilize the structure body array, each element in structure body array is the data of one structure body category.

The method for defining structure body array is similar to that of defining structure body variable, it is only necessary to define it as arrays. For example

```
struct Student            //Declaration structure body type
{
    int stu_ID;           //Student number
    char name[20];        //Student name
    float score ;         //Scores
};
struct Student stu[5];    //Defining structure body array
```

The defined structure body array stu contains a total of five elements, that is from stu[0] to stu[4], the elements in each array is of struct Student structural type.

6.2.6 结构体数组的初始化

对结构体数组,可以进行初始化赋值。例如:

```
struct Student
{
    int stu_ID; // 学号
    char name[20]; // 姓名
    float score; // 成绩
};
struct Student stu[3]={{11001, "Li ping" , 45},{1002, "Zhao min" ,62.5},{11003, "He fen" , 92.5}} ;
```

6.2.7 结构体数组的应用

【例 6.4】用结构体的方法实现多名同学基本信息的输入和输出。

【解题思路】先定义一个结构体,其成员有学号、姓名、三门课的成绩;再定义一个结构体数组并输入数组中每个元素的数值;最后将同学的信息输出。

6.2.6 The Initialization of Structure Body Array

For structure body array, the initialized assignment of value can be performed. For example

```
struct Student
{
    int stu_ID; //Student number
    char name[20]; //Name
    float score; //Scores
};
struct Student stu[3]={{11001, "Li ping",45},{1002, "Zhao min",62.5},{11003, "He fen",92.5}} ;
```

6.2.7 The Application of Structure Body Array

【Example 6.4】Applying the structure body method to realize the inputting and outputting of basic information for several students.

【Problem-Solving Ideas】First of all, defining a structure body, its members including student number, student name, and the scores of three subjects; Next, defining a structure array and input the values for each element in this array; At last, outputting the student information.

```c
1   #include "stdio.h"
2   #define N 3
3   void main()
4   {
5       int i;
6       //定义结构体类型
7       struct stu
8       {
9           char id[6];
10          char name[10];
11          int m1,m2,m3;
12      }student[N]; //定义结构体数组,共有N个元素,每个元素都具有struct stu的形式
13      for (i=0;i<N;i++)//输入结构体数组每个元素的信息
14      {
15          printf("请输入第%d个同学的记录:",i+1);
16          scanf("%s%s%d%d%d",student[i].id,&student[i].name,&student[i].m1,&student[i].m2,&student[i].m3);
17      }
18      printf("他们的成绩单为:\n");//输出结构体数组每个元素的信息
19      printf("学号\t姓名\t数学\t英语\t语文\n");
20      for(i=0;i<N;i++)
21          printf("%s\t%s\t%d\t%d\t%d\n",student[i].id,student[i].name,student[i].m1,student[i].m2,student[i].m3);
22  }
```

Sixth Task: Producing Students Report Cards

注意：对于【例 6.4】

（1）定义一个名为 stu 的结构体，其成员有 id、name、m1、m2、m3。

```
struct stu
{
    char id[6];
    char name[10];
    int m1,m2,m3;
};
```

（2）struct stu student[N]; 定义了一个名为 student 的结构体数组，共有 N 个元素，student[0]~ student[N-1] 每个元素都具有 struct stu 的结构形式，即具有相同的成员信息。对每个成员信息输入的时候应使用 for 循环语句实现，每个数组元素均需要输入 id、name、m1、m2、m3。

```
for (i=0;i<N;i++)
{
printf("请输入第 %d 个同学的记录:",i+1);
scanf("%s%s%d%d%d",student[i].id,&student[i].name,&student[i].m1,
&student[i].m2,&student[i].m3);
}
```

（3）对于结构体数组成员的输出也需要使用 for 循环语句实现，每个数组元素均需要输出 id、name、m1、m2、m3 信息。

```
for(i=0;i<N;i++)
printf("%s\t%s\t%d\t%d\t%d\n",student[i].id,student[i].name,
student[i].m1,student[i].m2,student[i].m3);
```

【例 6.5】用结构体的方法，计算三个同学的总成绩、平均成绩。

Attention: for【Example 6.4】

(1) Defining a structure body named stu, its members including id, name, m1, m2 and m3.

```
struct stu
{
    char id[6];
    char name[10];
    int m1,m2,m3;
};
```

(2)struct stu student[N]; Defining a structure body array named student, there is a total of N elements, for student[0] to student[N-1], each element possesses struct stu structural form, namely, possessing the same member information. When inputting the information of each member, it is necessary to apply for loop statement to realize it, the id, name, m1, m2 and m3 need to be input for each array element.

```
for (i=0;i<N;i++)
{
printf("Please input the record for the %dth student:",
i+1);
scanf("%s%s%d%d%d",student[i].id,&student[i].name,&student[i].m1,
&student[i].m2,&student[i].m3);
}
```

(3) The outputting for members in structure body array also requires the for loop statement to realize, the id, name, m1, m2 and m3 information all need to be output for each array element.

```
for(i=0;i<N;i++)
printf("%s\t%s\t%d\t%d\t%d\n",student[i].id,student[i].name,
student[i].m1,student[i].m2,student[i].m3);
```

【Example 6.5】By means of the method of structure body to calculate the total scores and average scores of three students.

【解题思路】先定义一个结构体，其成员有学号、姓名、三门课的成绩、总分和平均分；再定义一个结构体数组并输入数值；计算三个同学的总分和平均分并输出。

【Problem-Solving Ideas】First of all, defining a structure body, which is composed of students number, student name, the scores of three subjects, total scores and average scores; Next, define a structure body array and input the values; Calculating the total scores and average scores of three students and output.

```c
#include "stdio.h"
#define N 3
//定义结构体类型
struct stu
{
    char id[6];
    char name[10];
    int m1,m2,m3;
    float sum,avg;
};
void main()
{
    struct stu student[N];//定义结构体数组，共有N个元素，每个元素都具有struct stu的形式
    int i;
    for(i=0;i<N;i++)//输入结构体数组每个元素的数值，即三个同学的个人信息
    {
        printf("请输入第%d个同学的记录:",i+1);
        scanf("%s%s%d%d%d",student[i].id,&student[i].name,&student[i].m1,&student[i].m2,&student[i].m3);
    }
    for(i=0;i<N;i++)//计算每个同学的总分和平均分
    {
        student[i].sum=student[i].m1+student[i].m2+student[i].m3;
        student[i].avg=student[i].sum/3;
    }
    printf("他们的成绩单为:\n");//输出三个人的个人信息
    printf("学号\t姓名\t数学\t英语\t语文\t总分\t平均分\n");
    for(i=0;i<N;i++)
        printf("%s\t%s\t%d\t%d\t%d\t%.1f\t%.1f\n",student[i].id,student[i].name,student[i].m1,student[i].m2,student[i].m3,student[i].sum,student[i].avg);
}
```

6.3 任务实现

【任务要求】某班有40位学生参加了期终考试，考了三门课，分别是C语言、高等数学、英语，现要求输入该班同学的个人信息（包括学号、姓名、三门课的成绩），输出按照总分从高到低进行排序后的成绩单。

【任务分析】在本任务中将用结构体数组进行操作。具体步骤是：首先，进行学生信息的输入/输出；其次，计算每个同学的三门课的平均分；最后，按平均分

6.3 Task Implementation

【Task Requirements】There are forty students have participated in the final examinations in certain class, the examined subjects are C language, superior mathematics and English, now it is required to input the personal information of students in this class (including student number, student name and the scores for three subjects), outputting the transcript after ranking by descending order.

【Task Analysis】Operating the structure body array in this task. The specific procedures are: First of all, performing the outputting/inputting of student information; Secondly, calculating the average scores of three

Sixth Task: Producing Students Report Cards

的高低排序后输出成绩单。所以分成两个任务介绍：①用结构体数组进行学生信息的输入/输出；②输出排序后的学生成绩单。

6.3.1 学生信息的输入/输出

定义结构体 stu，其成员位学生的学号、姓名、三门课的成绩等个人信息；定义一个结构体数组 student[N]，进行多个学生信息的输入和输出。有两种实现方法：

（1）结构体在函数内。

subjects for each student; At last, outputting the transcript according to average scores by descending order. They will be divided into two tasks for introduction: ① performing the outputting/inputting of student information by means of structure body arrays; ② outputting the student transcripts after ranking.

6.3.1 The Input/Output of Student Information

Defining the structure body stu, the student number, student name, scores of three subjects and other personal information of member position; Defining one structure body array student[N], carrying out the outputting and inputting of information for several students. There are two implementation methods:

(1)The structure body is within function.

```
1  #include "stdio.h"
2  #define N 3
3  void main()
4  {
5      int i;
6      //定义结构体类型
7      struct stu
8      {
9          char id[6];
10         char name[10];
11         int m1,m2,m3;
12     }student[N]; //定义结构体数组，共有N个元素，每个元素都具有struct stu的形式
13     for (i=0;i<N;i++)//输入结构体数组每个元素的信息
14     {
15         printf("请输入第%d个同学的记录:",i+1);
16         scanf("%s%s%d%d%d",student[i].id,&student[i].name,&student[i].m1,&student[i].m2,&student[i].m3);
17     }
18     printf("他们的成绩单为:\n");//输出结构体数组每个元素的信息
19     printf("学号\t姓名\tC语言\t高等数学\t英语\n");
20     for(i=0;i<N;i++)
21     printf("%s\t%s\t%d\t%d\t%d\n",student[i].id,student[i].name,student[i].m1,student[i].m2,student[i].m3);
22  }
```

（2）结构体在函数外。

(2) The structure body is beyond function.

```
1   #include "stdio.h"
2   #define N 3
3   struct stu//定义结构体类型
4   {
5       char id[6];
6       char name[10];
7       int m1,m2,m3;
8   };
9   void main()
10  {
11      int i;
12      struct stu student[N];  //定义结构体数组,共有N个元素,每个元素都具有struct stu的形式
13      for (i=0;i<N;i++)//输入结构体数组每个元素的信息
14      {
15          printf("请输入第%d个同学的记录:",i+1);
16          scanf("%s%s%d%d%d",student[i].id,&student[i].name,&student[i].m1,&student[i].m2,&student[i].m3);
17      }
18      printf("他们的成绩单为:\n");//输出结构体数组每个元素的信息
19      printf("学号\t姓名\tC语言\t高数\t英语\n");
20      for(i=0;i<N;i++)
21          printf("%s\t%s\t%d\t%d\t%d\n",student[i].id,student[i].name,student[i].m1,student[i].m2,student[i].m3);
22  }
```

6.3.2 输出排序后的成绩单

为了方便运行调试,假设班上只有 5 名同学。定义一个能存放 40 个学生学号、姓名、数学、英语、语文、总分、平均分的结构体数组;从键盘输入 40 个同学的学号、姓名、三门课程信息;求每个同学的总分、平均分;对总分进行排序;输出排序后的成绩单。

6.3.2 Outputting the Transcript after Ranking

For the convenience of invoking, supposing that there are only five students in class. Define a structural array that can hold 40 students' student numbers, names, mathematies, English, Chinese, total scores and average scores.; Inputting the students number, name, and course information of three subjects of forty students from keyboard; Calculating the total scores and average scores of each student; Ranking the total scores; Outputting the transcript after ranking.

```
1   #include "stdio.h"
2   #define N 5
3   void main()
4   {
5       struct student//定义结构体类型
6       {
7           char id[6],name[10];
8           int m1,m2,m3;
9           float sum,avg;
10      }stu1[N],t;//定义结构体数组stu和结构体变量t
11      int i,j;
12      printf("请输入学生的信息: id、姓名、三门课成绩\n");
13      for(i=0;i<N;i++)//输入学生信息
14          scanf("%s%s%d%d%d",stu1[i].id,stu1[i].name,&stu1[i].m1,&stu1[i].m2,&stu1[i].m3);
15      for(i=0;i<N;i++)//计算学生三门课程总分和平均分
16      {
17          stu1[i].sum=stu1[i].m1+stu1[i].m2+stu1[i].m3;
18          stu1[i].avg=stu1[i].sum/3.0;
19      }
20      for(i=0;i<N-1;i++)//按照总分高低进行排序
21          for(j=0;j<N-1-i;j++)
22              if(stu1[j].sum<stu1[j+1].sum)
23              {
24                  t=stu1[j];
25                  stu1[j]=stu1[j+1];
26                  stu1[j+1]=t;
27              }
28      printf("排序后的成绩单为:\n");//输出排序后的成绩单
29      printf("学号\t姓名\tC语言\t高数\t英语\t总分\t平均分\n");
30      for(i=0;i<N;i++)
31          printf("%s\t%s\t%d\t%d\t%d\t%.1f\t%.1f\n",stu1[i].id,stu1[i].name,stu1[i].m1,
32              stu1[i].m2,stu1[i].m3,stu1[i].sum,stu1[i].avg);
33  }
```

Sixth Task: Producing Students Report Cards

【练习与提高】

1. 用户的信息，包括姓名、年龄、电话、籍贯，其信息分别为:{ "Liu" ,34, "5643213" , "Guangzhou" }、{ "Xu" , 27, "2113456" , "Shanghai" }、{ "Zhang" , 26, "2201100" , "Wuhan")、("Yang" .3 "6201101" , "Shenzhen" }, 请编程按照他们的姓名降序进行输出显示。

2. 某图书室购买了一批书，现编程要输入这批书的有关信息：书名、出版社、作者、定价。用结构体数组实现输入/输出图书的基本信息，输出时要求每行显示一本图书记录。

3. 某部门有职工 10 人，职工信息包括职工号、职工名、性别、年龄、工龄、工资和地址。通过键盘输入信息，输出工资/年龄的最大值和最小值的职工信息并按照工资高低进行排序。

【 Practice and Improvement 】

1. The information of users includes name, age, phone number and native place, they are {"Liu", 34, "5643213", Guangzhou"}, {"Xu", 27, "2113456", "Shanghai"}, {"Zhang", 26, "2201100", "Wuhan") and ("Yang". 3 "6201101", "Shenzhen"} respectively, please perform programming and output to display in accordance with their names by descending order.

2. A library has purchased a batch of books, currently, it is necessary to program and input the information about these books: book name, publishing house, author, and pricing. By means of structure body arrays to realize the outputting/inputting of basic information about these books, it is required to display one book record in each line when outputting.

3. There are ten personnel in one department, the personnel information includes personnel number, personnel name, gender, age, length of service, salary and address. Inputting the information by keyboard, outputting the personnel information with maximum and minimum salary/age, and ranking according to the salaries.

附录 A 运算符的优先级和结合性
Appendix A Priority and Associativity of Operators

优先级 priority	运算符 operators	解释 explanation	结合方向 binding direction
1	()	圆括号	由左向右
	[]	下标运算符	
	->	指向结构体成员运算符	
	.	结构体成员运算符	
2	!	逻辑非运算符	由右向左
	~	按位取反运算符	
	++	前缀增量运算符	
	--	前缀增量运算符	
	+	正号运算符	
	-	负号运算符	
	(类型)	类型转换运算符	
	*	指针运算符	
	&	地址与运算符	
	sizeof	长度运算符	
3	*	乘法运算符	由左向右
	/	除法运算符	
	%	取余运算符	
4	+	加法运算符	由左向右
	-	减法运算符	
5	<<	左移运算符	由左向右
	>>	右移运算符	
6	< <= > >=	关系运算符	由左向右

Appendix A   Priority and Associativity of Operators

续表

优先级 priority	运算符 operators	解释 explanation	结合方向 binding direction
7	==	等于运算符	由左向右
	!=	不等于运算符	
8	&	按位与运算符	由左向右
9	^	按位异或运算符	由左向右
10	|	按位或运算符	由左向右
11	&&	逻辑与运算符	由左向右
12	||	逻辑或运算符	由左向右
13	?:	条件运算符	由右向左
14	= += -= *= /= %= &= ^= |= <<= >>=	赋值运算符	由右向左
15	,	逗号运算符	由左向右

说明：同一优先级的运算符优先级别相同，运算次序由结合方向决定。

Explanation: Operators with the same priority have the same priority level, and the order of operations is determined by the combination direction.

附录 B ASCII 字符编码表
Appendix B ASCII Character Encoding Table

ASCII 值 ASCII value	控制字符 control character	ASCII 值 ASCII value	控制字符 control character	ASCII 值 ASCII value	控制字符 control character	ASCII 值 ASCII value	控制字符 control character
0	NUT	32	(space)	64	@	96	、
1	SOH	33	!	65	A	97	a
2	STX	34	"	66	B	98	b
3	ETX	35	#	67	C	99	c
4	EOT	36	$	68	D	100	d
5	ENQ	37	%	69	E	101	e
6	ACK	38	&	70	F	102	f
7	BEL	39	,	71	G	103	g
8	BS	40	(	72	H	104	h
9	HT	41	)	73	I	105	i
10	LF	42	*	74	J	106	j
11	VT	43	+	75	K	107	k
12	FF	44	,	76	L	108	l
13	CR	45	-	77	M	109	m
14	SO	46	.	78	N	110	n
15	SI	47	/	79	O	111	o
16	DLE	48	0	80	P	112	p
17	DCI	49	1	81	Q	113	q
18	DC2	50	2	82	R	114	r
19	DC3	51	3	83	X	115	s
20	DC4	52	4	84	T	116	t
21	NAK	53	5	85	U	117	u
22	SYN	54	6	86	V	118	v
23	TB	55	7	87	W	119	w

Appendix B ASCII Character Encoding Table

续表

ASCII 值 ASCII value	控制字符 control character	ASCII 值 ASCII value	控制字符 control character	ASCII 值 ASCII value	控制字符 control character	ASCII 值 ASCII value	控制字符 control character
24	CAN	56	8	88	X	120	x
25	EM	57	9	89	Y	121	y
26	SUB	58	:	90	Z	122	z
27	ESC	59	;	91	[	123	{
28	FS	60	<	92	\	124	|
29	GS	61	=	93	]	125	}
30	RS	62	>	94	^	126	~
31	US	63	?	95	—	127	DEL

附录 C 常用标准库函数

Appendix C Common Standard Library Functions

附表 C.1 数学标准库函数（math.h）Mathematical Standard Library Functions

函数名 function name	函数形式 functional form	功能 function	类型 type
abs	abs(int i)	求整数的绝对值	int
fabs	fabs(double x)	返回浮点数的绝对值	double
floor	floor(double x)	向下舍入	double
fmod	fmod(double x, double y)	计算 x 对 y 的模，即 x/y 的余数	double
exp	exp(double x)	指数函数	double
log	log(double x)	对数函数 ln(x)	double
log10	log10(double x)	对数函数 log	double
labs	labs(long n)	取长整型绝对值	long
modf	modf(double value, double *iptr)	把数分为指数和尾数	double
pow	pow(double x, double y)	指数函数 (x 的 y 次方)	double
sqrt	sqrt(double x)	计算平方根	double
sin	sin(double x)	正弦函数	double
asin	asin(double x)	反正弦函数	double
sinh	sinh(double x)	双曲正弦函数	double
cos	cos(double x);	余弦函数	double
acos	acos(double x)	反余弦函数	double
cosh	cosh(double x)	双曲余弦函数	double
tan	tan(double x)	正切函数	double
atan	atan(double x)	反正切函数	double
tanh	tanh(double x)	双曲正切函数	double

Appendix C　Common Standard Library Functions

附表 C.2　字符函数和字符串函数（string.h）Character functions and string functions

函数名 function name	函数形式 functional form	功能 function	类型 type
isalpha	isalpha(int ch)	假设 ch 是字母（'A' ～ 'Z','a' ～ 'z'）返回非 0 值，否则返回 0	int
isalnum	isalnum(int ch)	假设 ch 是字母（'A' ～ 'Z','a' ～ 'z'）或数字（'0' ～ '9'）返回非 0 值，否则返回 0	int
isascii	isascii(int ch)	假设 ch 是字符(ASCII 码中的 0 ～ 127)返回非 0 值，否则返回 0	int
istrl	istrl(int ch)	假设 ch 是作废字符 (0x7F) 或普通控制字符 (0x00 ～ 0x1F) 返回非 0 值，否则返回 0	int
isdigit	isdigit(int ch)	假设 ch 是数字（'0' ～ '9'）返回非 0 值，否则返回 0	int
isgraph	isgraph(int ch)	假设 ch 是可打印字符 (不含空格)(0x21 ～ 0x7E) 返回非 0 值，否则返回 0	int
islower	islower(int ch)	假设 ch 是小写字母（'a' ～ 'z'）返回非 0 值，否则返回 0	int
isprint	isprint(int ch)	假设 ch 是可打印字符 (含空格)(0x20 ～ 0x7E) 返回非 0 值，否则返回 0	int
ispunct	ispunct(int ch)	假设 ch 是标点字符 (0x00 ～ 0x1F) 返回非 0 值，否则返回 0	int
isspace	isspace(int ch)	假设 ch 是空格（' '），水平制表符（'\t'），回车符（'\r'），走纸换行（'\f'），垂直制表符（'\v'），换行符（'\n'），返回非 0 值，否则返回 0	int
isupper	isupper(int ch)	假设 ch 是大写字母（'A' ～ 'Z'）返回非 0 值，否则返回 0	int
isxdigit	isxdigit(int ch)	假设 ch 是 16 进制数（'0' ～ '9','A' ～ 'F','a' ～ 'f'）返回非 0 值，否则返回 0	int
tolower	tolower(int ch)	假设 ch 是大写字母（'A' ～ 'Z'）返回相应的小写字母（'a' ～ 'z'）	int
toupper	toupper(int ch)	假设 ch 是小写字母（'a' ～ 'z'）返回相应的大写字母（'A' ～ 'Z'）	int
strcat	strcat(char *dest,const char *src)	将字符串 src 添加到 dest 末尾	char

续表

函数名 function name	函数形式 functional form	功能 function	类型 type
strchr	strchr(const char *s,int c)	检索并返回字符 c 在字符串 s 中第一次出现的位置	char
strcmp	strcmp(const char *s1,const char *s2)	比较字符串 s1 与 s2 的大小，并返回 s1-s2	int
stpcpy	stpcpy(char *dest,const char *src)	将字符串 src 复制到 dest	char
strlen	strlen(const char *s)	返回字符串 s 的长度	int
strlwr	strlwr(char *s)	将字符串 s 中的大写字母全部转换成小写字母，并返回转换后的字符串	char

附表 C.3　输入输出函数（stdio.h）Input and output functions

函数名 function name	函数形式 functional form	功能 function	类型 type
getch	getch()	从控制台(键盘)读一个字符，不显示在屏幕上	int
putch	putch()	向控制台(键盘)写一个字符	int
getchar	getchar()	从控制台(键盘)读一个字符，显示在屏幕上	int
putchar	putchar()	向控制台(键盘)写一个字符	int
getchar	getchar()	从控制台(键盘)读一个字符，显示在屏幕上	int
getc	getc(FILE *stream)	从 stream 处读一个字符，并返回这个字符	int
putc	putc(int ch,FILE *stream)	向 stream 写入一个字符 ch	int
getw	getw(FILE *stream)	从 stream 读入一个整数，错误返回 EOF	int
putw	putw(int w,FILE *stream)	向 stream 写入一个整数	int
fclose	fclose(handle)	关闭 handle 所表示的文件处理	FILE *
fgetc	fgetc(FILE *stream)	从 stream 处读一个字符，并返回这个字符	int
fputc	fputc(int ch,FILE *stream)	将字符 ch 写入 stream 中	int
fgets	fgets(char *string,int n,FILE *stream)	从 stream 中读 n 个字符存入 string 中	char *

Appendix C   Common Standard Library Functions

续表

函数名 function name	函数形式 functional form	功能 function	类型 type
fopen	fopen(char *filename,char *type)	翻开一个文件 filename,翻开方式为 type,并返回这个文件指针,type 可为以下字符串加上后缀	FILE *
fputs	fputs(char *string,FILE *stream)	将字符串 string 写入 stream 中	int
fread	fread(void *ptr,int size,int nitems,FILE *stream)	从 stream 中读入 nitems 个长度为 size 的字符串存入 ptr 中	int
fwrite	fwrite(void *ptr,int size,int nitems,FILE *stream)	向 stream 中写入 nitems 个长度为 size 的字符串,字符串在 ptr 中	int
fscanf	fscanf(FILE *stream,char *format[,argument,…])	以格式化形式从 stream 中读入一个字符串	int
fprintf	fprintf(FILE *stream,char *format[,argument,…])	以格式化形式将一个字符串写给指定的 stream	int
scanf	scanf(char *format[,argument…])	从控制台读入一个字符串,分别对各个参数进展赋值,使用 BIOS 进展输出	int
printf	printf(char *format[,argument,…])	发送格式化字符串输出给控制台(显示器),使用 BIOS 进展输出	int